Stochastic Processes - Theoretical Advances and Applications in Complex Systems

Edited by Don Kulasiri

Published in London, United Kingdom

Stochastic Processes - Theoretical Advances and Applications in Complex Systems
http://dx.doi.org/10.5772/intechopen.1001590
Edited by Don Kulasiri

Contributors
Alexandra Pinto, Bruno Carbonaro, Dennis Quispe, Don Kulasiri, Federica Vitale, Hamilton C. Chinwenyi, Hussaini D. Ibrahim, Obidio Rubio, Parul Tiwari, Sandhya Samarasinghe, Theophilus Danjuma, Yeng Chai Soh, Yung Kuan Foo

First published in London, United Kingdom, 2024 by IntechOpen
IntechOpen is the global imprint of INTECHOPEN LIMITED, registered in England and Wales, registration number: 11086078, 167-169 Great Portland Street, London, W1W 5PF, United Kingdom

British Library Cataloguing-in-Publication Data
A catalogue record for this book is available from the British Library

Additional hard and PDF copies can be obtained from orders@intechopen.com

Stochastic Processes - Theoretical Advances and Applications in Complex Systems
Edited by Don Kulasiri
p. cm.
Print ISBN 978-1-83769-550-8
Online ISBN 978-1-83769-549-2
eBook (PDF) ISBN 978-1-83769-551-5

Meet the editor

Professor Don Kulasiri is a professor (personal chair) and has been head of the Centre for Advanced Computational Solutions (C-fACS) at Lincoln University since 1999. His research theme is to understand the mathematical basis for biological and environmental phenomena based on physics broadly construed to assimilate data-driven machine learning and AI in phenomenological models. He has been a visiting professor at the Mathematical Institute, Oxford University, the UK, since 2008; Princeton University, USA (2004, 2006); and the Mechanics and Computation Division, Stanford University, USA (1998) and a fellow of the New Zealand Centre at Peking University (2018) and of the Modelling and Simulation Society of Australia and New Zealand (MSSANZ). He has published over 190 publications and authored six research monographs on his research with leading international publishers.

For EU product safety concerns:
IN TECH d.o.o., Prolaz Marije Krucifikse Kozulić 3, 51000 Rijeka, Croatia,
info@intechopen.com or visit our website at intechopen.com.

Contents

Preface IX

Chapter 1 1
Introductory Chapter: Stochastic Processes – Theoretical Advances and Applications in Complex Systems
by Don Kulasiri

Chapter 2 5
Numerical Methods in Python for Solving Systems of Stochastic Differential Equations: An Application to the SIRD Model in Latin America
by Dennis Quispe and Obidio Rubio

Chapter 3 27
Perspective Chapter: Numerical Solutions for Modelling Complex Systems with Stochastic Differential and Partial Differential Equations (SDEs/SPDEs)
by Parul Tiwari, Don Kulasiri and Sandhya Samarasinghe

Chapter 4 53
Some Remarks on Vector Markov Chains and Their Applications to the Description of Many-Particle Systems
by Bruno Carbonaro and Federica Vitale

Chapter 5 71
Brownian Computing and Its Similarities with Our Brains
by Alexandra Pinto

Chapter 6 81
Martingales and Partial Differential Equations Price Valuation Models for European Put Options
by Hamilton C. Chinwenyi, Hussaini D. Ibrahim and Theophilus Danjuma

Chapter 7 107
Optimal Stochastic H^{∞} Controller Synthesis via LQR Theory
by Yung Kuan Foo and Yeng Chai Soh

Preface

Stochastic processes provide essential mathematical frameworks for various applications in complex systems that are better described using probability theory. Many phenomena and processes in life and medical sciences, physics, and engineering are subject to fluctuations due to the interactions with the environments within which they manifest. These fluctuations are often random and sometimes inherent to the phenomena, but they are also due to the measurements. While macroscopic systems can be successfully described by models rooted in Newtonian mechanics, mesoscopic and microscopic systems need probabilistic descriptions, and stochastic processes have been successful in providing the support to understand the complexity at these levels. At the macroscopic level, we can employ mechanistic models to explain the regularities in phenomena superimposing the randomness of variables. Randomness, however, affects the systems differently depending on the conditions under which they function. For example, noise-induced state space transitions are a well-known phenomenon in many natural systems. These can be dealt with using the theory of stochastic ordinary differential and partial differential equations.

Master equations are used in modeling mesoscopic systems such as the biomolecular interactions within a cell where you find experimentally established stochasticity due to low copy numbers and random collisions. But at microscopic levels, the quantum theory has been successful and is based on the uncertainty that exists within the system. At each level of natural systems, uncertainty plays a vital role; hence stochastic processes have become a cornerstone of the body of knowledge to understand complexity. Operational research is another discipline where stochastic processes such as Markov chains have been used for decades.

This book contains chapters written by colleagues who have been working with various complex systems. There is a broad span of topics covered by the authors, and I would like to thank them for their contributions. I would also like to thank IntechOpen for publishing this book; especially, I gratefully acknowledge Ms. Nina Miocevic, publishing process manager, for helping me in various ways in this project. Without her, this book project would not have been possible. I also thank Lincoln University for providing the facilities and the time to engage in editing this book.

Don Kulasiri
Centre for Advanced Computational Solutions (C-fACS),
Lincoln University,
Christchurch, New Zealand

Chapter 1

Introductory Chapter: Stochastic Processes – Theoretical Advances and Applications in Complex Systems

Don Kulasiri

1. Role of randomness in complex systems

Randomness is ubiquitous in complex systems. Many systems that we encounter in day-to-day life, whether in macroscopic or microscopic scales, exhibit random fluctuations caused by often interwoven internal and external causes. In the theory of biological evolution [1], random variation plays a pivotal role in selection. Dynamic systems are often modeled by deterministic differential equations, and these equations sometimes exhibit bifurcations with the time evolution of a critical parameter. Bifurcation points of the parameter divide the time evolution of the system along different branches [2]. Fluctuations in parameters near critical points can lead systems to bifurcations [3].

The concept of stability is extended to understand the dynamic nature of system behaviors when the parameters are perturbed under internal and external fluctuations. Self-organization of dissipative structures in irreversible reactions is a prime example of order when the variables are not in static states but in dynamic states. These self-organized behaviors are evident in open, far-from-thermodynamic equilibrium systems [4]. Beyond a certain threshold of fluctuations, chaos may exist. Hence to understand complex behaviors under fluctuation regimes, we need mathematical tools, theories, and frameworks. This is where a vast body of literature on stochastic processes based on probability theory can be very useful.

2. Stochastic processes

Mathematics of stochastic processes has been developed over almost 100 years with rigors of characterization of probability theory, and even though most of the stochastic processes are domains of mathematicians, there are many applications in engineering, physics, economics, and biology.

Markov processes are one of the most applied branches of stochastic processes that rely on the assumption (Markov property) that the current state of a system only depends on the previous state. This implies a state transition dynamism and the transition probabilities to guide the transition. The Markov property reduces the evolution of a complex system to a tractable mathematical framework. In other words,

IntechOpen

the Markov property is memoryless, limiting the memory only to the previous state. But many behaviors of complex systems such as weather/climate systems are based on many past events converging onto the present moment. Therefore, non-Markovian processes also should play a significant role in applications.

It is often important to understand the role of noise in system variables. Theories of stochastic differential equations (SDEs) may be applied in certain systems to explore the impact of noise on dependent variables. To this end, Ito calculus is a significant addition [5] to the theory. Many applications now exist using SDEs as well as Stochastic Partial Differential Equations (SPDEs). SPDEs are used in space-time problems with irregular properties in the medium in which the modeled phenomenon exists. The flow of solutes in porous media is a classic example [6].

Large-scale computations and numerical methodologies developed in recent decades have improved the applicability of SDEs/SPDEs on a more practical level. Indeed, this advancement has demonstrated the relevance of these equations to a wide range of technological disciplines, including reliability and control theory, data-driven computer prediction and design, decision-making, and risk analysis in the environmental and financial sectors. In the presence of inherent and extrinsic noise, SDEs/SPDEs are being utilized as a mathematical tool to describe numerous physical, biological, and economic systems. Modeling uncertainties, inbuilt parameters, and aspects of the applied theory such as open-end boundary conditions are all instances of intrinsic noise. In contrast, extrinsic noise can be caused by environmental factors, geographic disparity, or random user input. Another reason for using SPDEs relates to the need to find a controlled or adjustable solution to complicated physical systems. The advent of noise frequently leads to the emergence of a new phenomenon, both mathematically and phenomenologically.

There are literally hundreds of new advances in stochastic processes during the last several decades. Hence, readers are encouraged to understand the depth and breadth through a thorough literature review of the subject.

Author details

Don Kulasiri
Center for Advanced Computational Solutions, Lincoln University, Christchurch, New Zealand

*Address all correspondence to: don.kulasiri@lincoln.ac.nz

References

[1] Kirchner MW, Gerhart JC. The Plausibility of Life – Resolving Darwin's Dilemma. 1st ed. New Haven: Yale University Press; 2005. 352 p

[2] Hazewinkel M, Jurkovich R, Paelinck JHP, editors. Bifurcation Analysis: Principles, Applications and Synthesis. Dordrecht, Holland: D. Reidel Publishing Company; 1984. 484 p. DOI: 10.1007/978-94-009-6239-2

[3] Horsthemke W, Lefever R. Noise-Induced Transitions: Theory and Applications in Physics, Chemistry, and Biology. Berlin, Heidelberg: Springer-Verlag; 2006. 318 p

[4] Nicolis G, Prigogine I. Self-Organization in Nonequilibrium Systems: From Dissipative Structures to Order through Fluctuations. New York, London, Sydney, Toronto: John Wiley & Sons; 1977. 491 p

[5] Oksendal B. Stochastic Differential Equations: An Introduction with Applications. 5th ed. Heidelberg, New York: Springer-Verlag; 2000. 330 p

[6] Kulasiri D. Non-Fickian Solute Transport in Porous Media: A Mechanistic and Stochastic Theory. Heidelberg, New York, Dordrecht, London: Springer; 2013. 230 p. DOI: 10.1007/978-3-642-34985-0

Chapter 2

Numerical Methods in Python for Solving Systems of Stochastic Differential Equations: An Application to the SIRD Model in Latin America

Dennis Quispe and Obidio Rubio

Abstract

In this chapter, we apply the exploration of the Euler–Maruyama, Milstein, and Runge–Kutta methods to solve systems of stochastic differential equations associated with the stochastic SIRD model. We simulate sample trajectories of each variable using Python and data collected in Peru during the years 2020 and 2021, marked by the onset of the COVID-19 pandemic. Our research involves comparing stochastic and deterministic systems obtained from the SIRD model in the Peruvian context. We use different values for the intensity of white noise to assess the impact of stochasticity on the dynamics of the SIRD model. Presenting random simulations alongside deterministic ones provides a comprehensive understanding of the effect of randomness in the context of infectious disease modeling.

Keywords: SIRD model, stochastic differential equations, numerical methods, Euler–Maruyama method, Milstein method, Runge–Kutta method, Python code

1. Introduction

In our investigation into the dynamics of infectious diseases, we leverage advanced numerical methods to address the complexities posed by systems governed by stochastic differential equations (SDE). Our primary focus is on the widely recognized SIRD (Susceptible-Infectious-Recovered-Deceased) model, particularly relevant in the Latin American context.

Python serves as our computational tool of choice, ensuring the efficient implementation of numerical methods. Our study is grounded in data collected in Peru during 2020 and 2021, a period marked by the profound impact of the COVID-19 pandemic. This temporal context provides a unique opportunity for a meticulous examination of the influence of stochasticity on the SIRD model, as evidenced by our previous works [1, 2].

Our methodology involves simulating trajectories of SIRD variables in Python, considering varying intensities of white noise resembling Brownian motion. We employ the Euler–Maruyama, Milstein, and Runge–Kutta methods, each strategically chosen for its precision and relevance to infectious disease dynamics within the framework of stochastic epidemiological systems.

In the mathematical modeling of COVID-19 transmission, our study aligns with prior research [1], which utilized the deterministic SIR model for COVID-19 transmission in Peru. In parallel, our investigation [2] employs the stochastic SIRD model, utilizing the Milstein method to simulate and demonstrate results with Peruvian data. This approach promises a nuanced exploration of SIRD dynamics under diverse conditions, dissecting the interplay between deterministic trends and stochastic fluctuations. Our work contributes valuable insights to the field of infectious disease dynamics, shedding light on the complexities inherent in real-world scenarios.

2. SIRD model

The SIRD model, representing Susceptible-Infectious-Recovered-Deceased, offers a theoretical foundation for capturing the dynamics of infectious diseases in a population. This model categorizes individuals into four compartments based on their health status concerning the specific disease.

1. *Susceptible (S)*: Individuals in this group are susceptible to the infectious agent but have not yet been exposed or infected. The number of susceptibles diminishes as individuals encounter the disease.

2. *Infectious (I)*: This compartment includes individuals currently infected and capable of transmitting the disease to susceptible individuals. The duration of infectivity significantly influences disease spread.

3. *Recovered (R)*: Individuals in this category have recovered from the infection and are assumed to have acquired immunity. Recovered individuals are no longer susceptible or infectious.

4. *Deceased (D)*: Unfortunately, some individuals may succumb to the disease, transitioning to the deceased category. The mortality rate is a crucial parameter governing this transition.

The SIRD model's dynamics are commonly expressed through a system of differential equations, where transition rates between compartments depend on parameters such as transmission, recovery, and mortality rates. This model provides valuable insights into the evolution of infectious diseases within a population, aiding researchers and policymakers in assessing intervention impact and disease control strategies.

In epidemiology, the SIRD model stands as a fundamental tool, offering a quantitative understanding of interactions among population compartments during infectious disease outbreaks. Its widespread use is attributed to its versatility and simplicity, making it a valuable framework for analyzing and predicting the spread of infectious diseases.

2.1 Deterministic model

In a given population N, if there are individuals affected by a virus, the population is categorized into the infected group (I), the susceptible group (S), the recovered group (R), and the deceased group (D). These groups undergo changes over time following the depicted scheme in **Figure 1**.

Expressed as a modification of the mathematical model proposed in 1927 by Kermack and McKendrick [3], it is represented by the system of four equations:

$$\begin{cases} S'(t) &= -\lambda S(t)I(t), \\ I'(t) &= \lambda S(t)I(t) - \gamma I(t) - \mu I(t), \\ R'(t) &= \gamma I(t), \\ D'(t) &= \mu I(t). \end{cases} \tag{1}$$

Referred to as the SIRD model, the variables and parameters are detailed in **Table 1**.

All variables are required to meet the compatibility condition $N = S(t) + I(t) + R(t) + D(t)$, ensuring the constancy of the total population.

2.2 Stochastic model

Given the significant impact of environmental factors on COVID-19 spread, coupled with the inherent uncertainty in data accuracy, the stochastic aspect becomes particularly intriguing. The complexity is heightened by scenarios where infected individuals, especially children, may not show symptoms but contribute to transmission. Additionally, asymptomatic individuals, unmonitored yet active in community transmission, add another layer of complexity.

Transmission through contact with contaminated surfaces further widens the scope of possibilities. Acknowledging literature on modeling environmental fluctuations' influence on population dynamics [4, 5], two approaches emerge. One involves introducing random perturbations for each variable, delving into the less-understood connections among susceptible, infected, and recovered individuals through white noise. Following [6], a stochastic system is proposed as follows:

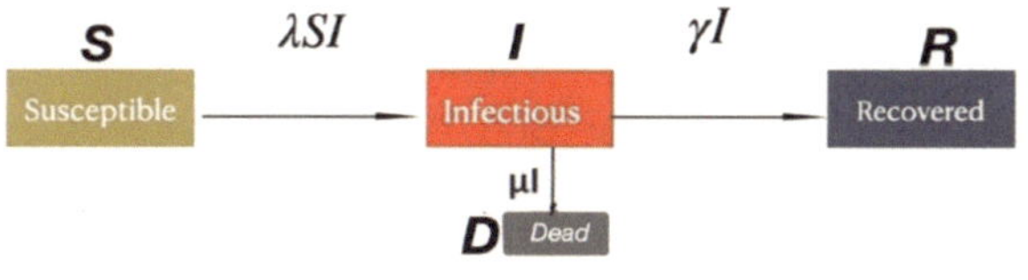

Figure 1.
The progression of a pandemic's development over time.

Variables and parameters	Meaning
$S(t)$	Susceptible population at time t,
$I(t)$	Infected population at time t,
$R(t)$	Recovered population at time t,
$D(t)$	Dead population at time t,
t	The time expressed in days,
λ	Transmission rate,
γ	Rate of recovery,
μ	Mortality rate.

Table 1.
Variables and parameters associated with the system (1).

$$\begin{cases} dS(t) &= -\lambda S(t)I(t)dt + \sigma_1 S(t)dW_1(t), \quad S(0) = s_0, \\ dI(t) &= (\lambda S(t)I(t)dt - \gamma I(t) - \mu I(t))dt + \sigma_2 I(t)dW_2(t), \quad I(0) = i_0, \\ dR(t) &= \gamma I(t)dt + \sigma_3 R(t)dW_3(t), \quad R(0) = r_0, \\ dD(t) &= \mu I(t)dt + \sigma_4 D(t)dW_4(t), \quad D(0) = d_0. \end{cases} \tag{2}$$

Here, $s_0, i_0, r_0, d_0 \geq 0$; $W_i(t)$ are independent standard Brownian motions, and $\sigma_i^2 \geq 0$ represent the intensities of for i = 1, 2, 3, 4.

The entire system is established on a comprehensive probability space $(\Omega, \mathcal{F}, P)$ accompanied by a filtration $\mathcal{F}_t \geq 0$ that adheres to standard conditions (i.e., it is right-continuous and increasing, with $\mathcal{F}_0$ encompassing all null sets on P).

3. Numerical methods

Stochastic differential equations (SDEs) serve the purpose of describing phenomena influenced by both deterministic and random terms.

$$\begin{cases} dX(t) = f(X(t))\,dt + g(X(t))\,dW(t), \quad t \geq 0, \\ X(0) = X_0. \end{cases} \tag{3}$$

Eq. (3) consists of a right-hand side depending on two components, the function $f : \mathbb{R}^d \to \mathbb{R}^d$, recognized as the drift, representing the deterministic aspect and the function $g : \mathbb{R}^d \to \mathbb{R}^{d\times m}$, referred to as the diffusion, governing the stochastic part.

The term $W(t)$ in (3) denotes an m-dimensional standard Wiener process. Because of its lack of differentiability everywhere (with probability 1), the expression in Eq. (3) serves as a concise notation for its integral form:

$$X(t) = X(0) + \int_0^t f(X(s))\,ds + \int_0^t g(X(s))\,dW(s). \tag{4}$$

If the stochastic integral in (4) is an Itô integral, the equation is an Itô stochastic differential equation. If it is a Stratonovich integral, (4) represents a Stratonovich stochastic differential equation. Unless stated otherwise, our discussion primarily revolves around Itô SDEs. Notably, Itô SDEs can be equivalently formulated in Stratonovich form:

$$dX(t) = \left(f(t) - \frac{1}{2}g'(t)\right) dt + g(t)\circ dW(t). \tag{5}$$

This interchangeability allows transitioning between Itô and Stratonovich SDEs, providing equivalent formulations in terms of the solution [7, 8].

3.1 Euler: Maruyama

The Euler–Maruyama (EM) method is recognized as one of the simplest computational techniques for approximating Stochastic Differential Equations (SDEs), serving as the stochastic counterpart to the Euler method employed in ordinary differential equations. Before delving into the stochastic scenario, let us provide a brief introduction to this method.

To approximate the solution of a Stochastic Differential Equation (SDE) expressed in the form (3) over the interval [0, T], the interval is initially discretized into L parts ($L \in \mathbb{N}$). Here, $\Delta t = T/L$ represents the time step size, and $\tau_j = j\Delta t$, $j = 0, \ldots, L$. Our numerical approximation is denoted as X_j. The Euler–Maruyama method, when applied to an SDE of the type (3), is precisely defined by the following recursive formula:

$$X_{j+1} = X_j + f(X_j)\Delta t + g(X_j)\Delta W_{j+1}, \quad j = 0, 1, \ldots, L-1. \tag{6}$$

In this context, we compute the discretized trajectories of the Brownian motion and use them to generate the increments $\Delta W_{j+1} = W(\tau_{j+1}) - W(\tau_j)$ needed in the above recursive formula. For convenience, we always choose the time step $\Delta t = \tau_{j+1} - \tau_j$ as an integer multiple $R \geq 1$ of the increment $\delta = t_{j+1} - t_j$ for the trajectory that generates the Brownian sample, that is, $\Delta t = R\delta t$. This ensures that the set of points $\{\tau_j\}$ in each discretized Brownian trajectory contains the points $\{\tau_j\}$ in each calculated solution of the Euler–Maruyama method. The increments $W(\tau_{j+1}) - W(\tau_j)$ are given by:

$$W(\tau_{j+1}) - W(\tau_j) = W((j+1)R\delta t) - W(jR\delta t) = \sum_{k=jR+1}^{(j+1)R} \delta W_k. \tag{7}$$

Finally, we describe the Euler–Maruyama method as follows:
Euler–Maruyama Method (3.1):

$$\begin{cases} X_0 &= x_0, \quad \Delta t = R\delta t = R(t_{j+1} - t_j) = \tau_{j+1} - \tau_j, \\ X_{j+1} &= X_j + f(X_j)\Delta t + g(X_j)\{W(\tau_{j+1}) - W(\tau_j)\}, \; j = 0, 1, \ldots, L-1. \end{cases} \tag{8}$$

The method mentioned above is of the explicit one-step type since each value X_{j+1} is expressed in terms of a single previous value, X_j. The denomination

Euler–Maruyama method is then the summa of the work of Euler (1707–1783) and Gisiro Maruyama (1916–1986). Subsequently, the Euler–Maruyama method emerges as a first-order truncation of the Itô-Taylor expansion of the precise solution to (3).

3.2 Milstein method

Additional formulae for approximating SDEs (3) may arise by including further terms from the Itô-Taylor expansion in the numerical method. An instance of this is the Milstein method, which, for scalar problems, can be obtained as follows. Applying the Itô formula to the expressions $f(X_s)$ and $g(X_s)$, where these represent the coefficients of the Stochastic Differential Equation (SDE), we can derive

$$X_{t_{j+1}} = X_{t_j} + \int_{t_j}^{t_{j+1}} (f\left(X_{t_j}\right) + \int_{t_j}^{s} \left(f'(X_u)f(X_u) + \frac{1}{2}f''(X_u)g(X_u)^2\right) du + \int_{t_j}^{s} f'(X_u)g(X_u)dW_u)\, ds + \int_{t_j}^{t_{j+1}} (g\left(X_{t_j}\right) + \int_{t_j}^{s} \left(g'(X_u)f(X_u) + \frac{1}{2}g''(X_u)g(X_u)^2\right) du + \int_{t_j}^{s} g'(X_u)g(X_u)\, dW_u)\, dW_s. \tag{9}$$

We can ignore the double integrals of the form $dW_s du$ and $ds\, du$. Then, we obtain the following expression:

$$X_{t_{j+1}} \approx X_{t_j} + \int_{t_j}^{t_{j+1}} f\left(X_{t_j}\right) ds + \int_{t_j}^{t_{j+1}} \left(g\left(X_{t_j}\right) + \int_{t_j}^{s} g'(X_u)g(X_u)\, dW_u\right) dW_s. \tag{10}$$

This can be expressed as:

$$X_{t_{j+1}} \approx X_{t_j} + f\left(X_{t_j}\right)\Delta t + g\left(X_{t_j}\right)\Delta W_{j+1} + \int_{t_j}^{t_{j+1}} \int_{t_j}^{s} g'(X_u)g(X_u)\, dW_u\, dW_s.$$

The first three terms are familiar from the Euler–Maruyama method. The fourth term can be approximated as:

$$\int_{t_j}^{t_{j+1}} \int_{t_j}^{s} g'(X_u)g(X_u)\, dW_u\, dW_s \approx g'\left(X_{t_j}\right)g\left(X_{t_j}\right) \int_{t_j}^{t_{j+1}} \int_{t_j}^{s} dW_u dW_s.$$

The integral on the right-hand side is known to be:

$$\int_{t_j}^{t_{j+1}} \int_{t_j}^{s} dW_u dW_s = \frac{1}{2}\left(\left(\Delta W_{j+1}\right)^2 - \Delta t\right)$$

Substituting into our previous approximation, we finally obtain the Milstein method *Milstein Method (3.2)*:

$$\begin{cases} X_0 = x_0 \\ X_{j+1} = X_j + f(X_j)\Delta t + g(X_j)\Delta W_{j+1} + \frac{1}{2}g'(X_j)g(X_j)\left((\Delta W_{j+1})^2 - \Delta t\right) \end{cases} \tag{11}$$

while for systems of SDEs,

$$dX(t) = g^0(X(t))\Delta t + \sum_{k=1}^{m} g^k(X(t))\, dW^k(t),$$

where $g^k : \mathbb{R}^d \to \mathbb{R}^d$, , $k = 0, 1, \dots, m$, the method is given by

$$\begin{aligned} X_{n+1} = X_n + g^0(X_n)\Delta t + \sum_{k=1}^{m} g^k(X_n)\Delta W_{n+1}^k \\ + \sum_{j_1, j_2 = 1}^{m} L^{j_1} g^{j_2}(X_n) \int_{t_n}^{t_{n+1}} \int_{t_n}^{t} dW^{j_1}(s)\, dW^{j_2}(t), \end{aligned}$$

with $L^k = \sum_{i=1}^{d} g^{k,i} \frac{\partial}{\partial x_i}$, $k = 1, 2, \dots, m$.

The Milstein method, also a one-step explicit approach, achieves increased accuracy compared to the Euler–Maruyama method due to the inclusion of additional terms originating from the truncation of the Itô-Taylor expansion.

3.3 Runge–Kutta

A category of stochastic Runge–Kutta methods (SRK methods) has been derived by appropriately perturbing deterministic Runge–Kutta methods. The formulation we adopt is based on the framework outlined in [9], which characterizes SRK methods as follows:

$$X_{n+1} = X_n + \Delta t \sum_{i=1}^{s} b_i f(\hat{X}_i) + \Delta W_{n+1} \sum_{i=1}^{s} q_i g(\hat{X}_i), \tag{12}$$

Where the intermediate values are used to estimate the value of $X(t_n + c_i \Delta t)$ for every index $i = 1, 2, \dots, s$, are provided by

$$\hat{X}_i = X_n + \Delta t \sum_{j=1}^{s} a_{ij} f(\hat{X}_j) + \Delta W_{n+1} \sum_{j=1}^{s} \gamma_{ij} g(\hat{X}_j), \tag{13}$$

A more compact effective representation of SRK methods (12)–(13) is obtained by using a Butcher tableau:

$$\begin{array}{c|cccc|cccc} c_1 & a_{11} & a_{12} & \cdots & a_{1s} & \gamma_{11} & \gamma_{12} & \cdots & \gamma_{1s} \\ c_2 & a_{21} & a_{22} & \cdots & a_{2s} & \gamma_{21} & \gamma_{22} & \cdots & \gamma_{2s} \\ \vdots & \vdots & \vdots & \ddots & \vdots & \vdots & \vdots & \ddots & \vdots \\ c_s & a_{s1} & a_{s2} & \cdots & a_{ss} & \gamma_{s1} & \gamma_{s2} & \cdots & \gamma_{ss} \\ \hline & b_1 & b_2 & \cdots & b_s & q_1 & q_2 & \cdots & q_s \end{array}$$

with $\mathbf{A} = (\mathbf{a_{i,j}}), \Gamma = (\gamma_{i,j}) \in \mathbb{R}^{s \times s}$, and vector of weights $\mathbf{b} = (\mathbf{b_j})$, $\mathbf{q} = (\mathbf{q_j}) \in \mathbb{R}^s$ and vector of abscissae $\mathbf{c} = (\mathbf{c_j}) \in \mathbb{R}^s$.

If Γ is the zero matrix and $\mathbf{q}$ the null vector, then SRK methods (12)–(13) recovering the equivalent family of deterministic Runge–Kutta methods is also applicable. In the stochastic scenario, the computational cost of the method is influenced by the composition of matrices in (13). Specifically, if $\mathbf{A}$ and Γ are strictly lower triangular, the corresponding method is explicit.

3.4 Convergence analysis

Convergence in the stochastic context differs due to the randomness of the error, denoted as $\|X_n - X(t_n)\|$, subject to random fluctuations. The error is measured using the expectation operator.

Assuming analysis within the interval $[0, T]$, partition it into L subintervals, with $\Delta t = \frac{T}{L}$. The partition intersects a set of grid points $\mathcal{I}_{\Delta t} = \{t_j = j\Delta t | j = 0, 1, \dots, L\}$. Typically, these points are a subset of Wiener points, $\Delta t = R\delta t, R \in \mathbb{N}$.

Definition 1: A numerical method for SDEs (3) providing the numerical solution $\{X_n\}_{n=0}$ is strongly convergent if

$$\lim_{\Delta t \to 0} \sup_{t_n \in \mathcal{I}_{\Delta t}} \mathbb{E}\left[|X_n - X(t_n)|\,|\right] = 0. \tag{14}$$

Moreover, we say that its strong order of convergence is p if there exist two positive numbers C and Δt^* such that

$$\sup_{t_n \in \mathcal{I}_{\Delta t}} \mathbb{E}\left[|X_n - X(t_n)|\,|\right] \le C\Delta t^p, \quad \text{for any } \Delta t \le \Delta t^*. \tag{15}$$

It is noteworthy to mention that in numerous real-world scenarios, an integer p that satisfies (15) may not be present, particularly in cases involving low regularity problems, such as certain stochastic partial differential equations. An alternative concept of convergence is presented as follows:

Definition 2: A numerical method for SDEs (3) providing the numerical solution is weakly convergent if

$$\lim_{\Delta t \to 0} \sup_{t_n \in \mathcal{I}_{\Delta t}} |\mathbb{E}[\Phi(X_n)] - \mathbb{E}[\Phi(X(t_n))]| = 0 \tag{16}$$

for any test function Φ within a designated function space S. Furthermore, we assert that its weak order of convergence is p if there exist two positive numbers C and Δt^*, satisfying

$$\sup_{t_n \in \mathcal{I}_{\Delta t}} |\mathbb{E}\left[\Phi(X)\right] - \mathbb{E}\left[\Phi(X(t_n))\right]| \le C\Delta t^p, \quad \text{for any } \Delta t \le \Delta t^*. \tag{17}$$

In numerous real-world scenarios, a suitable choice for the functional space S is given by the space of algebraic polynomials limited to a specified degree. It is crucial to underscore the connection between the two previously mentioned convergence concepts. Assuming that $\Phi(x) = x$, a method demonstrating strong convergence is inherently weakly convergent.

$$\sup_{t_n \in \mathcal{I}_{\Delta t}} |\mathbb{E}\,[\Phi(X_n)] - \mathbb{E}\,[\Phi(X(t_n))]| = \sup_{t_n \in \mathcal{I}_{\Delta t}} |\mathbb{E}\,[X_n] - \mathbb{E}\,[X(t_n)]| \tag{18}$$

Since

$$\sup_{t_n \in \mathcal{I}_{\Delta t}} |\mathbb{E}\,[X_n] - \mathbb{E}\,[X(t_n)]| = \sup_{t_n \in \mathcal{I}_{\Delta t}} |\mathbb{E}\,[X_n - X(t_n)]| \leq \sup_{t_n \in \mathcal{I}_{\Delta t}} \mathbb{E}\,[\|X_n - X(t_n)\|], \tag{19}$$

We obtain

$$\sup_{t_n \in \mathcal{I}_{\Delta t}} |\mathbb{E}\,[\Phi(X)] - \mathbb{E}\,[\Phi(X(t_n))]\;| \leq \sup_{t_n \in \mathcal{I}_{\Delta t}} \mathbb{E}\,[\|X_n - X(t_n)\|], \tag{20}$$

as we aimed to prove.

Below, we present some results on the convergence of the numerical methods discussed in this section. The proofs are omitted and can be found in [9].

Theorem 1: Euler–Maruyama method (8) is strongly convergent with a strong order of $\frac{1}{2}$.

Theorem 2: Euler–Maruyama method (8) is weakly convergent with weak order 1.

It is also possible to demonstrate the following, [9]:

- Both the strong and weak orders of convergence for the Milstein method (10) are equivalent and set at 1.
- Consequently, the Milstein method represents an improvement over the Euler–Maruyama method, particularly in terms of strong convergence.
- The convergence of explicit SRK methods (12)–(13) relies on the convergence of the corresponding deterministic RK methods, subject to an additional condition. This additional condition is necessary to ensure that

$$\sum_{i=1}^{s} b_i = \sum_{i=1}^{s} q_i = 1.$$

4. An application to the case of Peru

4.1 Data and equations

Based on the national census conducted on October 22, 2017, the population of Peru is reported to be 31,237,385 individuals. Subsequent yearly estimates, as of June 30, according to [10], project the population to approximately N = 32,625,948, as indicated by this institution.

To apply the system of eqs. (1) for parameter calculation, we require the values of $S(t)$ $I(t)$, $R(t)$, and $D(t)$. These values have been derived from official reports of the Ministry of Health (MINSA) [11] spanning from March 5 to September 30, 2020. A subset of this data is presented in **Table 2** for illustrative purposes.

In order to calculate the parameters, we use the approximation of the derivative by central difference, implying that for each day of the sample data $t = 1, 2, \dots, 208 = T$, we calculate the parameters λ_t, γ_t, and μ_t, which give us a sequence of data. Then, we calculate the simple average of such values to have the final parameters, as shown in **Table 3**.

Day	Date	Susceptible $S(t)$	Infected $I(t)$	Recovered $R(t)$	Deceased $D(t)$
0	Mar-05	32,625,948	0	0	0
1	Mar-06	32,625,947	1	0	0
12	Mar-17	32,625,831	116	0	1
26	Mar-31	32,624,883	630	394	41
27	Apr-01	32,624,625	829	447	47
33	Apr-07	32,622,994	1734	1100	120
40	Apr-14	32,615,645	7204	2869	230
47	Apr-21	32,608,111	10,371	6982	484
56	Apr-30	32,588,972	25,520	10,405	1051
57	May-01	32,585,489	28,206	11,129	1124
71	May-15	32,541,453	54,956	27,147	2392
87	Jun-01	32,455,909	96,898	68,507	4634
117	Jul-01	32,337,471	100,372	178,245	9860
148	Aug-01	32,203,765	111,940	290,835	19,408
179	Sept-01	31,962,511	154,001	480,177	29,259
208	Sept-30	31,807,651	95,234	690,528	32,535

Table 2.
Data related to COVID-19 in Peru (mar. 05 to sept. 30, 2020).

t = 0	S(0)	I(0)	R(0)	D(0)	λ	γ	μ
March 17	32,625,831	116	0	1	4.244×10^{-9}	4.576×10^{-2}	1.990×10^{-3}

Table 3.
Parameters of the SIR model for COVID-19 in Peru (mar. 17 to sept. 30, 2020).

To obtain an approximate solution for the system (2), we consider $\Delta W_l \sim \zeta_l \sqrt{dt}$, where ζ_l is a standard Gaussian random variable with mean 0 and variance 1. Subsequently, we discretize the system using the Euler–Maruyama, Milstein's, and Runge–Kutta methods, as outlined in (8), (10), (12), and (13). The Python codes obtained for both the deterministic case and the stochastic system are presented in the following section.

4.2 Codes in python and simulations

Below, we present the complete codes for both deterministic and stochastic SIRD models.

4.2.1 Deterministic case

```
import numpy as np.
import matplotlib.pyplot as plt.
# Model parameters.
```

```
T0 = 0, T = 365, n = 36,500, dt = (T - T0) /n.
lambda_val = 0.00000000424404.
gamma = 0.0457600061.
mu = 0.0019902810.
initial_conditions = np.array ([32,625,831, 116, 0, 1]).
# Define the system of differential equations.
def sird_model (S, I, R, D):
dS = – lambda_val * S * I * dt.
dI = (lambda_val * S * I – gamma * I – mu * I) * dt.
dR = gamma * I * dt.
dD = mu * I * dt.
return np.array([dS, dI, dR, dD]).
# Initialize matrices to store results.
population_states = np.zeros((n, 4)).
population_states[0,:] = initial_conditions.
# Simulation of the model.
for i in range(1, n):
d_population = sird_model(*population_states[i – 1,:])
population_states[i,:] = population_states[i – 1,:] + d_population.
lot(np.arange(T0, T, dt), population_states[:, 0], label = 'Susceptible').
plt.plot(np.arange(T0, T, dt), population_states[:, 1], label = 'Infected').
plt.plot(np.arange(T0, T, dt), population_states[:, 2], label = 'Recovered').
plt.plot (np.arange(T0, T, dt), population_states [:, 3], label = 'Deceased').
plt.lege.
# Plotting results.
plt.pnd().
plt.grid(True).
plt.xlabel('Time').
plt.ylabel('Population').
plt.title('SIRD Model Simulation').
plt.show().
```

We observe in **Figure 2** that the simulation of the deterministic case using Python reproduces the same patterns obtained in MATLAB, which aligns with the literature [1, 2].

4.2.2 Euler–Maruyama code

```
import numpy as np.
import matplotlib.pyplot as plt.
# Model parameters.
T0 = 0, T = 365, n = 36,500, dt = (T - T0) / n.
lambda_val = 0.00000000424404.
gamma = 0.0457600061.
mu = 0.0019902810.
sigma1 = 0.05.
sigma2 = 0.05.
sigma3 = 0.05.
sigma4 = 0.05.
initial_conditions = np.array([32,625,831, 116, 0, 1]).
# Stochastic SIRD Model using Euler-Maruyama.
```

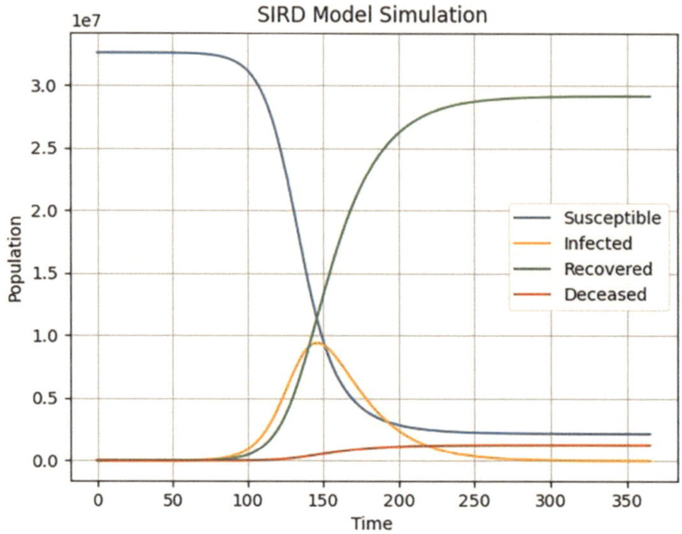

Figure 2.
Simulation results of the SIRD model in the deterministic case. The plot shows the evolution of the susceptible, infected, recovered, and deceased populations over time.

```
np.random.seed(100).
S = np.zeros(n).
I = np.zeros(n).
R = np.zeros(n).
D = np.zeros(n).
dW = np.sqrt(dt) * np.random.randn(n, 4).
S[0] = 32,625,831.
I[0] = 116.
R[0] = 0.
D[0] = 1.
def f(S, I, R, D, lambda_val, gamma, mu):
dS = −lambda_val * S * I.
dI = lambda_val * S * I – gamma * I – mu * I.
dR = gamma * I.
dD = mu * I.
return np.array([dS, dI, dR, dD]).
def g(X, sigma1, sigma2, sigma3, sigma4):
return np.array([[sigma1 * X[0], 0, 0, 0],
                 [0, sigma2 * X[1], 0, 0],
                 [0, 0, sigma3 * X[2], 0],
                 [0, 0, 0, sigma4 * X[3]]])

for i in range (1, n):
# Euler-Maruyama.
X = np.array([S[i – 1], I[i – 1], R[i – 1], D[i – 1]]).
dX = f(*X, lambda_val, gamma, mu) * dt +.
g(X, sigma1, sigma2, sigma3, sigma4).dot(dW[i,:])
    X_new = X + dX.
    S[i], I[i], R[i], D[i] = X_new.
```

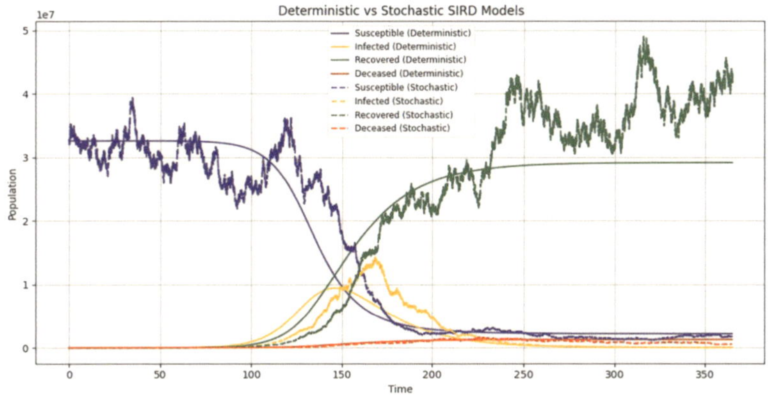

Figure 3.
Simulation results of the SIRD model using Euler–Maruyama method.

```
# Plotting results.
plt.figure (figsize = (12, 6)).
# Stochastic SIRD using Euler-Maruyama.
plt.plot(np.arange (0, T + dt, dt), [S[0]] + list (S),
label = 'Susceptible (Stochastic)', linestyle = 'dashed', color = 'blue').
plt.plot(np.arange(0, T + dt, dt), [I[0]] + list(I),
label = 'Infected (Stochastic)', linestyle = 'dashed', color = 'orange').
plt.plot(np.arange(0, T + dt, dt), [R[0]] + list(R),
label = 'Recovered (Stochastic)', linestyle = 'dashed', color = 'green').
plt.plot(np.arange(0, T + dt, dt), [D[0]] + list(D),
label = 'Deceased (Stochastic)', linestyle = 'dashed', color = 'red').
plt.legend(fontsize = 'small').
plt.grid(True).
plt.xlabel('Time').
plt.ylabel('Population').
plt.title('Deterministic vs. Euler-Maruyama').
plt.show().
```

We can observe in **Figure 3** that by using the values of $\sigma_1 = \sigma_2 = \sigma_3 = \sigma_4 = 0.05$, the curve of recovered individuals overestimates that of the deterministic case. This phenomenon is evident when conducting an individual variable analysis in **Figure 4** and when performing 50 simulations in **Figure 5**. We attribute this to the convergence order of the method, $p = 0.5$.

4.2.3 Milstein code

```
import numpy as np.
import matplotlib.pyplot as plt.
# Model parameters.
T0 = 0, T = 365, n = 36,500, dt = (T - T0) / n.
lambda_val = 0.00000000424404.
gamma = 0.0457600061.
mu = 0.0019902810.
sigma1 = 0.05, sigma2 = 0.05, sigma3 = 0.05, sigma4 = 0.05.
# Milstein method for the SIRD model.
```

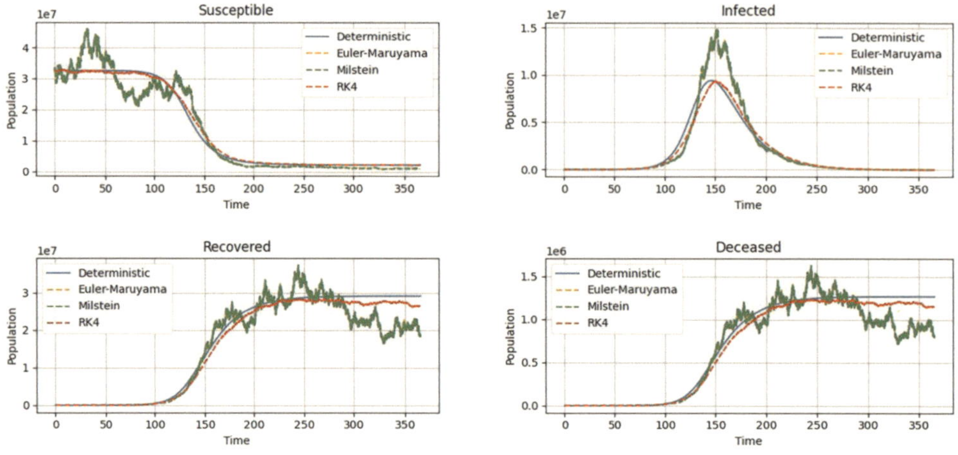

Figure 4.
Plots of the join three methods.

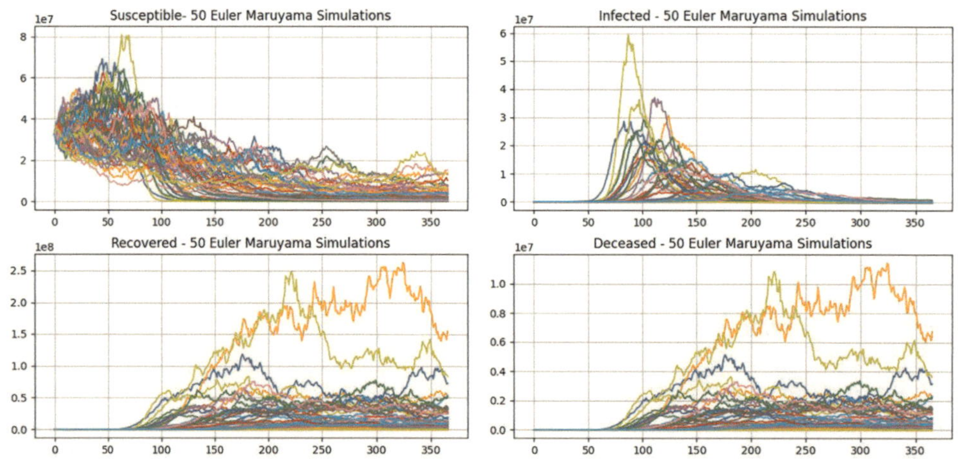

Figure 5.
50 simulations of each variable of the stochastic SIRD model using Euler–Maruyama method.

```
S_milstein = np.zeros(n).
I_milstein = np.zeros(n).
R_milstein = np.zeros(n).
D_milstein = np.zeros(n).
dW = np.zeros(n).
S_milstein[0] = 32,625,831.
I_milstein[0] = 116.
R_milstein[0] = 0.
D_milstein[0] = 1.
```

```
dW[0] = np.sqrt(dt) * np.random.randn().
for i in range(1, n):
dW[i] = np.sqrt(dt) * np.random.randn().
S_milstein[i] = S_milstein[i - 1] - lambda_val * S_milstein[i - 1].
* I_milstein[i - 1] * dt + sigma1 * S_milstein[i - 1]* dW[i - 1].
+ 0.5 * (sigma1 ** 2) * S_milstein[i - 1] * (dW[i - 1] ** 2 - dt).
I_milstein[i] = I_milstein[i - 1] + (lambda_val * S_milstein[i - 1].
* I_milstein[i - 1] - gamma * I_milstein[i - 1] - mu * I_milstein[i - 1]).
+ sigma2 * I_milstein[i - 1] * dW[i - 1] + 0.5 * (sigma2 ** 2).
* I_milstein[i - 1] * (dW[i - 1] ** 2 - dt).
R_milstein[i] = R_milstein[i - 1] + gamma * I_milstein[i - 1] * dt.
+ sigma3 * R_milstein[i - 1] * dW[i - 1] + 0.5 * (sigma3 ** 2).
* R_milstein[i - 1] * (dW[i - 1] ** 2 - dt).
D_milstein[i] = D_milstein[i - 1] + mu * I_milstein[i - 1] * dt +.
sigma4 * D_milstein[i - 1] * dW[i - 1] + 0.5 * (sigma4 ** 2).
* D_milstein[i - 1] * (dW[i - 1] ** 2 - dt).
# Plotting results.
plt.figure(figsize = (12, 6)).
plt.plot(np.arange(T0, T, dt), [S_milstein[0]] + list(S_milstein[1:]),
linestyle = 'dashed', label = 'Susceptible (Milstein)').
plt.plot(np.arange(T0, T, dt), [I_milstein[0]] + list(I_milstein[1:]),
linestyle = 'dashed', label = 'Infected (Milstein)').
plt.plot(np.arange(T0, T, dt), [R_milstein[0]] + list(R_milstein[1:]),
linestyle = 'dashed', label = 'Recovered (Milstein)').
plt.plot(np.arange(T0, T, dt), [D_milstein[0]] + list(D_milstein[1:]),
linestyle = 'dashed', label = 'Deceased (Milstein)').
plt.legend(fontsize = 'small').
plt.grid(True).
plt.xlabel('Time').
plt.ylabel('Population').
plt.title('SIRD Model Simulation with Deterministic and Milstein Methods').
plt.show().
```

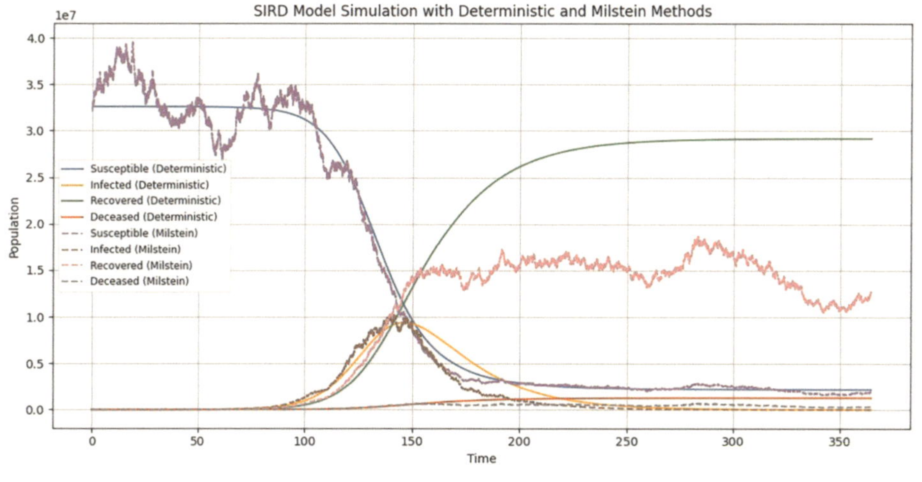

Figure 6.
Simulation results of the SIRD model using Milstein method.

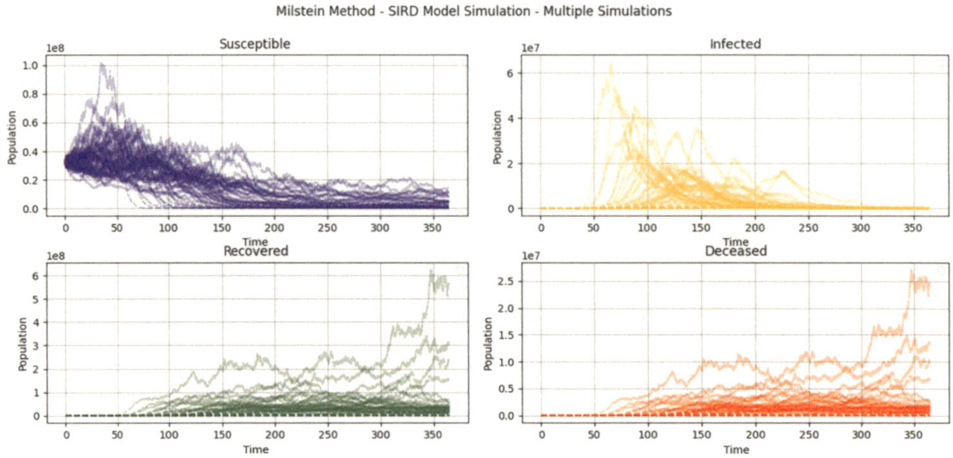

Figure 7.
50 simulations of each variable of the stochastic SIRD model using Milstein method.

In the case of Milstein with the same values for $\sigma_1 = \sigma_2 = \sigma_3 = \sigma_4 = 0.05$, we observe in **Figure 6** an underestimation in the recovered individuals. In the individual analysis depicted in **Figure 4**, it is noted that during a segment of its trajectory, an overestimation occurs, while in another segment, an underestimation is observed. Finally, in **Figure 7**, conducting 50 simulations reveals that, collectively, the trajectories are underestimated compared to the deterministic case. We attribute this phenomenon to the inherent convergence order of the method, $p = 1$.

4.2.4 Runge: Kutta 4 code

```
import numpy as np.
import matplotlib.pyplot as plt.
# Set random seed for reproducibility.
np.random.seed(100).
# Model parameters for the SIRD system.
T0 = 0, T = 365, n = 36,500, dt = (T - T0) / n.
lambda_val = 0.00000000424404.
gamma = 0.0457600061.
mu = 0.0019902810.
sigma1 = 0.05, sigma2 = 0.05, sigma3 = 0.05, sigma4 = 0.05.
# Initialize arrays to store S, I, R, D.
S = np.zeros(n).
I = np.zeros(n).
R = np.zeros(n).
D = np.zeros(n).
# Initialize stochastic increments.
dW = np.sqrt(dt) * np.random.randn(4, n) # Now dW has 4 components.
# Initial conditions.
S[0] = 32,625,831, I[0] = 116, R[0] = 0, D[0] = 1.
# Define deterministic part of the system.
def deterministic_part(S, I, R, D):
```

```
        return np.array([−lambda_val * S * I,
                         lambda_val * S * I - gamma * I - mu * I,
                         gamma * I,
                         mu * I]).
# Define stochastic part of the system.
def stochastic_part(S, I, R, D):
return np.array([sigma1 * S, sigma2 * I, sigma3 * R, sigma4 * D]).
# Stochastic Runge–Kutta method.
def stochastic_runge_kutta(X, f, g, dt, dW):
    stages = 4.
    A = np.array([[0, 0, 0, 0],
                  [0.5, 0, 0, 0],
                  [0, 0.5, 0, 0],
                  [0, 0, 1, 0]])
    Gamma = np.array([[0, 0, 0, 0],
                      [0.5, 0, 0, 0],
                      [0, 0.5, 0, 0],
                      [0, 0, 1, 0]])
    b = np.array([1 / 6, 1 / 3, 1 / 3, 1 / 6])
    q = np.array([1/6, 1/3, 1/3, 1/6]).
    c = np.array([0, 0.5, 0.5, 1]).
    X_hat = np.zeros((stages, len(X))).
    for i in range(stages):
        sum_f = sum(A[i, j] * f(*X_hat[j]) for j in range(stages)).
        sum_g = sum(Gamma[i, j] * g(*X_hat[j]) for j in range(stages)).
        X_hat[i] = X + dt * sum_f + dW[i] * np.sqrt(dt) * sum_g.
    sum_b = sum(b[j] * f(*X_hat[j]) for j in range(stages)).
    sum_q = sum(q[j] * g(*X_hat[j]) for j in range(stages)).
    X_new = X + dt * sum_b + dW[−1] * np.sqrt(dt) * sum_q.
    return X_new.
# Time-stepping loop.
for i in range(1, n):
    X = np.array([S[i - 1], I[i - 1], R[i - 1], D[i - 1]]).
    X_new = stochastic_runge_kutta(X, deterministic_part, stochastic_part,
      dt, dW[:, i - 1]).
    S[i], I[i], R[i], D[i] = X_new.
# Plot results.
plt.plot(np.arange(0, T + dt, dt), [S[0]] + list(S), label = 'Susceptible').
plt.plot(np.arange(0, T + dt, dt), [I[0]] + list(I), label = 'Infected').
plt.plot(np.arange(0, T + dt, dt), [R[0]] + list(R), label = 'Recovered').
plt.plot(np.arange(0, T + dt, dt), [D[0]] + list(D), label = 'Deceased').
plt.legend().
plt.grid(True).
plt.xlabel('Time').
plt.ylabel('Population').
plt.title('Stochastic Runge–Kutta Method for SIRD Model').
plt.show().
```

With the same values of $\sigma_1 = \sigma_2 = \sigma_3 = \sigma_4 = 0.05$, in this case, we observe in **Figure 8** that all four variables follow the deterministic path. This observation holds true both in the individual analysis depicted in **Figure 4** and in the 50 simulations

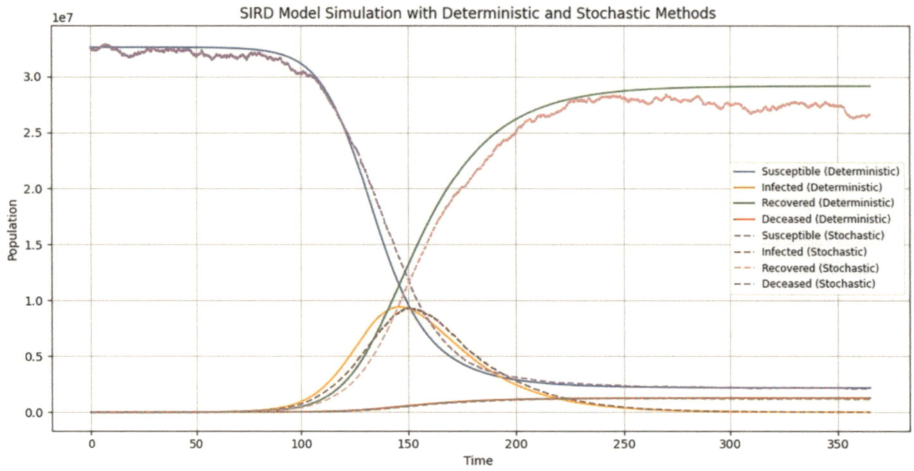

Figure 8.
Simulation results of the SIRD model using Runge Kutta 4 stochastic method.

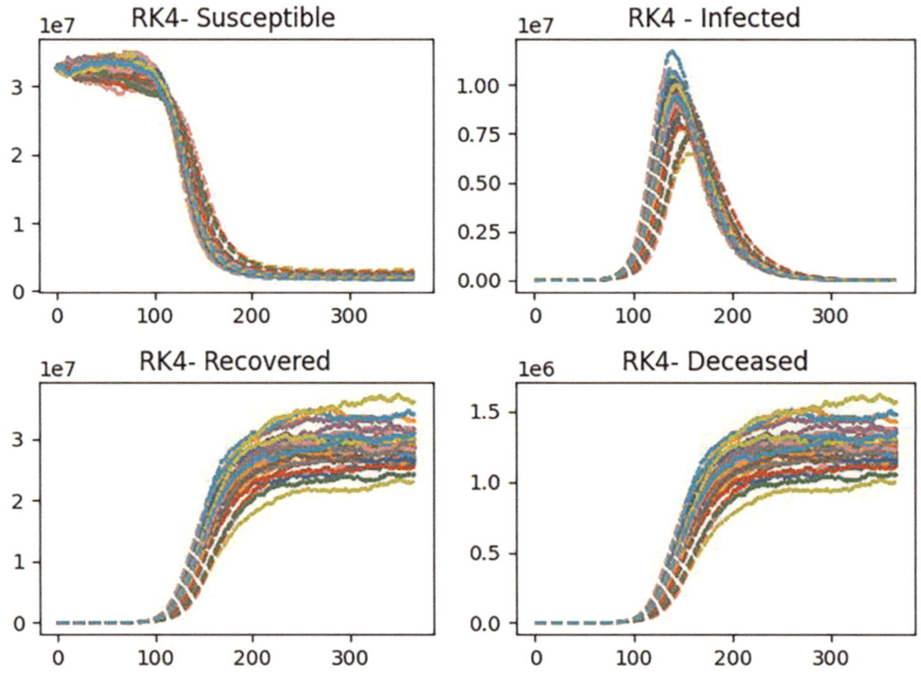

Figure 9.
50 simulations of each variable of the stochastic SIRD model using RK4 method.

presented in **Figure 9**. Analyzing the three methods, we can conclude that the fourth-order Runge–Kutta method best fits the deterministic case. We attribute this to the stability and convergence order of the method, $p = 1.5$, [7].

In **Figure 4**, we present a comparison for each variable of the stochastic SIRD model using the three methods explained in this chapter: Euler–Maruyama (yellow dashed lines), Milstein (green dashed lines), Runge–Kutta (red dashed lines), and the deterministic simulation (solid blue line).

In **Figures 5–9**, we conducted 50 simulations for each method to examine the average behavior of the simulations. It is worth noting that in some cases, stochastic models may provide a better description of the variable over extended periods, as seen in the case of recovered, for example.

5. Conclusion

5.1 Theoretical insights and stochastic methods

This section delves into theoretical results for stochastic methods, including discussions on the strong and weak convergence of the Euler–Maruyama and Milstein methods. Additionally, it explores specific conditions for the convergence of SRK methods, contingent on the deterministic Runge–Kutta methods' convergence.

5.2 Application to COVID-19 modeling in Peru

The study applies the developed methodologies to a specific case: modeling the spread of COVID-19 in Peru. Official health reports are utilized to calculate parameters for the SIR model. The section presents deterministic and stochastic simulations of the SIRD model, offering insights into population dynamics over time.

5.3 Implementation and comparative analysis

Providing practical implementation, this part offers Python code snippets for various stochastic methods, including Euler–Maruyama, Milstein, and stochastic Runge–Kutta. The concluding part includes comparative visualizations between deterministic and stochastic simulations, emphasizing the impact of stochastic components on the model's outcomes.

This comprehensive summary covers theoretical foundations, practical applications in COVID-19 modeling, and detailed code snippets, offering a thorough overview of the research.

Conflict of interest

The authors declare no conflict of interest.

Author details

Dennis Quispe* and Obidio Rubio†
National University of Trujillo, Trujillo, Peru

*Address all correspondence to: dquispe@unitru.edu.pe

† These authors contributed equally.

References

[1] Vergara E et al. Basic SIR epidemiological model for COVID-19: Case of the regions of Peru. Selecciones Matematicas. 2020;7(1):151-161. DOI: 10.17268/sel.mat.2020.01.14

[2] Rubio O et al. Simulation of the behavior of COVID-19 with the stochastic SIRD model: Case of Peru. Selecciones Matematicas. 2023;**10**(1): 81-89. DOI: 10.17268/sel.mat.2023.01.08

[3] Kermack WO, McKendrick AG. A contribution to the mathematical theory of epidemics. Proceedings of the Royal Society of London. 1927;**115**:700-721. Available from: https://royalsocietypublishing.org/doi/pdf/10.1098/rspa.1927.0118

[4] Baoa J, Maob X, Yinc G, Yuana C. Competitive Lotka–Volterra population dynamics with jumps. Nonlinear Analysis. 2011;**74**:6601-6616. DOI: 10.1016/j.na.2011.06.043

[5] Zhou Y, Zhang W, Yuan S. Survival and stationary distribution of an SIR epidemic model with stochastic perturbations. Applied Mathematics and Computation. 2014;**244**:118-131. DOI: 10.1016/j.amc.2014.06.100

[6] El Koufi A, El Koufi N. Stochastic differential equation model of COVID-19: Case study of Pakistan. Results in Physics. 2022;**34**:105218. DOI: 10.1016/j.rinp.2022.105218

[7] Waszkiewicz R, Bartczak M. Pychastic: Precise Brownian dynamics using Taylor-Ito integrators in python. SciPost Physics. 2023;**11**:1-20. DOI: 10.21468/SciPostPhysCodeb.11

[8] Huang C. Python solver for stochastic differential equations. Journal of the Mathematical Society of the Philippines. 2011;**34**:87-97

[9] Thomas. G Introduction to Stochastic Differential Equations. New York, NY: Marcel Dekker Inc; 1988. pp. 183-217

[10] INEI. PERU, Estimates and Projections of Population by Department, Province, and District, 2018–2020. Digital ed. INEI; 2020. Available from: https://proyectos.inei.gob.pe/Est/Lib0846

[11] Ministry of Health of Peru. Epidemiological Alert: Code: AE-015-2020. Available from: https://www.gob.pe/institucion/minsa/informes-publicaciones/1471880-alerta-epidemiologica-n-21-coronavirus-covid-19

Chapter 3

Perspective Chapter: Numerical Solutions for Modelling Complex Systems with Stochastic Differential and Partial Differential Equations (SDEs/SPDEs)

Parul Tiwari, Don Kulasiri and Sandhya Samarasinghe

Abstract

Understanding phenomena ranging from biological processes to financial markets involves uncertainty. Stochastic Differential Equations (SDEs) and Stochastic Partial Differential Equations (SPDEs) serve as robust mathematical frameworks for modelling such systems. Given the stochastic influences within these models, comprehending the dynamics of complex systems becomes pivotal for accurately predicting system behaviour. However, traditional numerical techniques frequently encounter challenges in effectively addressing the intricacies and stochastic properties inherent in these equations. This chapter explores several numerical methods that offer streamlined and dependable solutions capable of handling the complexities inherent in stochastic differential and partial differential equations. Also, numerical challenges associated with these methods are discussed and the solution strategies are also suggested.

Keywords: stochastic differential and partial differential equations, Brownian motion, Cameron-Martin basis, stochastic spectral methods, Wiener chaos expansion

1. Introduction

Mathematical modelling of complex system is a sophisticated way to understand and simulate intricate relationships in various fields including engineering, science, economics, finance and biology [1–4]. Stochastic Differential Equations (SDEs) and Stochastic Partial Differential Equations (SPDEs) play critical role in understanding the behaviour of these systems. These systems having dynamic interactions and a large number of variables, necessitating advanced mathematical and computational models for proper representation. To capture the complexities and uncertainties inherent in these systems, researchers employ mathematical frameworks using SDEs and SPDEs [5, 6].

Many physical processes are simulated using mathematical models that lead to partial differential deterministic equations. Due to lack of data on parameters and initial data, a system's behaviour may differ from the ideal deterministic description. Adding random inputs, such as random variables or stochastic processes, to make a system's explanation more realistic is one way to deal with this issue and compensate for a lack of information. As a result of this, SDEs and SPDEs are formulated.

In financial mathematics, these equations are a central part of modelling; these are used to model the key features such as interest rate, asset prices, risks and their volatility [7–9]. In applied mathematics, these equations are useful for modelling transport equations [10], quantum mechanics [11], wave propagation in random media [12], neuroscience [13], population genetics and mathematical biology [14]. These equations are also being used in pure mathematics as a natural object to study stochastic processes and dynamical systems and to develop theoretical frameworks, convergence and stability for the algorithms in the form of lemma and theorems.

Although numerous studies have proved the applicability of solute transport models based on deterministic advection-dispersion equations [15, 16], the success of such deterministic models in describing solute movement in variable field conditions is questionable. The deterministic modelling of solute transport in groundwater contamination involves the simulation of hydrodynamic dispersion, advection and absorption with velocity field. Moreover, these components suffer from significant uncertainty because of scale dependency, even in homogeneous media; their effects are poorer in the case of heterogeneous mediums where micro-variations take place in flow parameters.

These difficulties have been captured and handled in two ways. One way is to increase the dispersion parameter with distance, and the other way is to increase it with time. Another approach to scale dependency and heterogeneity is stochastic modelling in which relevant flow parameters are characterised by a distribution function [17]. Ensemble means of the output variables provide a useful description of the larger-scale spatial and temporal trends in practical applications. Although it is not of great importance to know the exact deviation of local fluctuations from the ensemble mean, it is useful to know their likely range.

Various numerical approaches are used to address the challenges posed by the dynamic and uncertain nature of complex systems. Finite difference approaches, for example, discretize the spatial domain and use difference approximations to obtain solutions [18]. This method works well in situations when the spatial features of a complicated system are carefully considered. Monte Carlo simulations, on the other hand, use random sampling to generate numerical results, making them ideal for capturing the randomness found in many complex systems [19, 20]. Efficiency of numerical methods is often measured in terms of computational resources and accuracy. Some numerical methods, like the Implicit-Explicit (IMEX) schemes, strike a balance between stability and efficiency, making them suitable for solving stiff systems with varying time scales. Karhunen-Loève expansion utilises eigenfunctions to represent stochastic processes. This approach is advantageous in problems with random spatial variations, allowing for a reduction of dimensionality.

Another numerical approach to solve SDEs and SPDEs is to use stochastic spectral methods (SSMs). These methods provide a powerful and flexible way for dealing with complex systems defined by these equations and use spectral representations to efficiently handle the systems' inherent stochasticity, offering correct solutions while maintaining computational efficiency.

Stochastic spectral methods are intrusive and non-intrusive. Intrusive methods [21, 22] represent the solution as a sum of orthogonal basis functions, and then using these basis functions to approximate the solution to the SDE or SPDE. The coefficients of the expansion are determined by solving a system of equations derived from the original problem, typically using techniques such as Galerkin projection or least-squares regression. One popular intrusive stochastic spectral method is the stochastic finite element method (SFEM). SFEM involves representing the solution as a sum of finite element basis functions, and then using these basis functions to approximate the solution [23, 24].

Non-intrusive spectral methods approximate the solution to the SPDE using a series of basis functions. The coefficients of the basis functions are determined by solving a system of equations derived from the original SPDE. These methods are 'non-intrusive' in the sense that they do not require the original deterministic solver to be modified in order to handle stochastic inputs. Instead, they use a surrogate model to represent the solution in terms of a spectral expansion. A popular non-intrusive stochastic spectral method is the polynomial chaos expansion (PCE) method. PCE involves representing the solution as a sum of orthogonal polynomials, such as Legendre or Hermite polynomials. The coefficients of the expansion are determined by projecting the original problem onto the polynomial basis, which can be done using techniques such as Galerkin projection or least-squares regression [25–27]. An advantage of non-intrusive stochastic spectral methods is their ability to handle high-dimensional problems with a relatively small number of samples. This is because the spectral expansion can capture the global behaviour of the solution, allowing for accurate approximation of the solution using only a few terms.

There is a great need and interest in developing efficient numerical schemes to solve these equations. It is essential and profitable too, especially when dealing with randomness. This chapter explores the existing numerical approaches surrounding SDEs and SPDEs relevant to selected applications. Computational aspects of different stochastic spectral methods are discussed and applied to solve some practical problems such as contaminant transport in porous media. A class of problems of solving SDEs/SPDEs using polynomial chaos expansion methods is discussed. The computational efficiencies of polynomial chaos expansion methods are compared with alternative methods. This chapter also explores Wick calculus, a potent but less commonly used method in the stochastic calculus.

2. Numerical approaches for SDEs and SPDEs

2.1 Direct integration techniques

In stochastic analysis, quantities of interest can be represented by the expected value of certain functionals of the dependent variable, say, y, and denoted as E. The value of this expectation is defined in the form of integrals and the need to compute certain integrals, with respect to the probability measure P_ξ as:

$$E(f) = \int_R f(y(x);x)dP_\xi(x) = \int_R f(y(x);x)p_\xi(x)dx \tag{1}$$

where $p(\xi)$ is the probability density function of ξ, and the integration strategies lead to the following estimation:

$$E(f) \approx Q_i(f) = \sum_{i=1}^{n} f(y(x_i); x_i) w_i \tag{2}$$

where the x_i are the points and $w_i \in R$ are the weights of integration. The model response is then evaluated for possible outcomes $\xi = x_i$ of the selected random variables. Finally, it requires the solution of a system having n uncoupled deterministic equations.

$$L(y(x_i); x_i) = b(x_i), i = 1, 2, \dots \dots \dots n \tag{3}$$

2.2 Finite difference methods

Finite difference methods involve discretizing the spatial and temporal domains of the equations. For instance, the Euler-Maruyama method is widely used for SDEs. In this approach, the stochastic process is simulated by iteratively updating the solution at each time step.

Consider the example of modelling stock prices S_t as given in Eq. (4), where the geometric Brownian motion equation can be solved using the Euler-Maruyama method. This numerical scheme is the generalisation of Euler's method for ODEs to solve SODEs. Consider an Itô SDE [28] of the form,

$$dS_t = \mu(t, S_t)dt + \sigma(t, S_t)dW_t \tag{4}$$

where W_t is a standard Wiener process on the interval $[0, T]$. To solve this equation numerically, we discretise the time interval with an h (a sufficiently small) step size. Symbolically, Eq. (4) can be written as,

$$S_t = S_{t_0} + \int_{t_0}^{t} \mu(t, S_t)dt + \int_{t_0}^{t} \sigma(t, S_t)dW_t \tag{5}$$

and the Euler-Maruyama (E-M) approximation is calculated using the Itô-Taylor expansion.

The first integral on the right-hand side of (Eq. (5)) using an Itô-Taylor expansion is,

$$\int_{t_0}^{t} \mu(t, S_t)dt = \int_{t_0}^{t} \mu(t_0 + (t - t_0), S_0 + S(t) - S_0)dt \tag{6}$$

$$= h\mu(t_0, S_0) + O\left(h^{3/2}\right) \tag{7}$$

or simply,

$$S_{n+1} = S_n + \mu(t_n, S_n)\int_{t_n}^{t_{n+1}} d\tau + \sigma(t_n, S_n)\int_{t_n}^{t_{n+1}} dW_\tau \tag{8}$$

where $n = 0, 1, \dots \dots \dots, N - 1$ for $t_0 < t_1 < \dots < t = T$ and $N \in N$. Thus, for an m-dimensional SDE, $S^h = \left(S_1^h, S_2^h, \dots, S_m^h\right)$ and

$$S^h_{t_{n+1}} = S^h_{t_n} + h\mu\left(t_n, S^h_{t_n}\right) + \sigma\left(t_n, S^h_{t_n}\right)\Delta B_n \tag{9}$$

with $\Delta B_n = B_{t_{n+1}} - B_{t_n}, t_n = nh$

ΔB_n is the sequence of identically independent Gaussian random variables having a zero mean and h standard deviation.

Eq. (9) can be rewritten as,

$$S_{t_{n+1}} - S_{t_n} = h\mu\left(t_n, S^h_{t_n}\right) + \sqrt{h}\sigma\left(t_n, S^h_{t_n}\right)B_n + O(h) \tag{10}$$

The order condition for h judges the quality of the numerical scheme. Consequently, the variation in step size h defines the convergence rate of the numerical scheme. In numerical approximation, based on model requirements, the two most - commonly used definitions of convergence are strong and weak convergence.

2.2.1 Numerical example

We shall apply the E-M method to a simple population growth model which has an exact solution. The population growth model equation of a species (S) in time t years is written as:

$$dS_t = \alpha S_t dt + \beta S_t dW_t \tag{11}$$

where α and β are the model parameters. The exact solution of this equation, using Itô's formula, is given by:

$$S_t = S_0 e^{\left(\left(\alpha - \frac{\beta^2}{2}\right)t + \beta W_t\right)} \tag{12}$$

Simulation (three runs) of the population growth model for an exact solution, with parameter values $\alpha = 0.35$ and $\beta = 0.25$, is shown in **Figure 1**. Initial population is assumed to be $S_0 = 20$ and $\Delta t = 0.01$. The E-M approximation is also plotted for the three runs for comparison in **Figure 2** with the same parameter values for all computations.

The simulation results are different for the same parameter values because of the stochastic nature of the model. Thus, it is a good idea to run the simulation numerous times and choose an average value to see how the given system behaves (**Figure 3**).

The first graph on the top left corner is the average population growth after five simulations. The results from the E-M scheme differ from the exact solution. However, as the simulation runs are increased, the E-M approximations coincide with the actual solution.

2.3 Monte Carlo (MC) methods

Monte Carlo methods are very popular for handling both SDEs and SPDEs. These methods involve random sampling to approximate solutions. MC methods approximate the outcomes based on random sampling and the law of inertia of large numbers. An example is the Black-Scholes model in finance, where Monte Carlo simulations can estimate option prices by simulating the stochastic evolution of asset prices. Practically, these random samples are pseudo-random samples of the basic

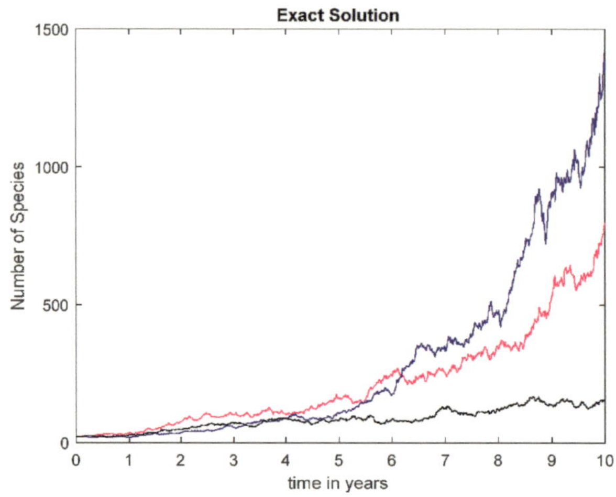

Figure 1.
Exact solution for population growth model.

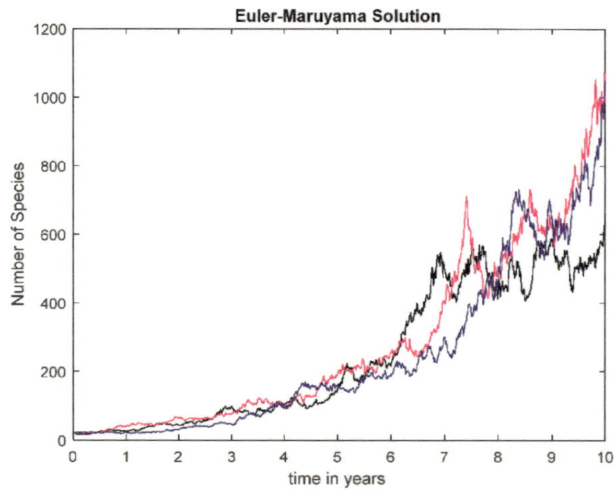

Figure 2.
Euler-Maruyama approximation for population growth model.

random variable ξ and integration error is defined by a Gaussian Random Variable (GRV) (**Figure 4**).

A basic MC estimator of an integral of a quantity of interest (expectation), say μ, is given by,

$$\text{Mean}(\mu) = \int_R f(x)p(x)dx = E[f(x)] \approx \frac{1}{n}\sum_{i=1}^{n} f(x_i), \tag{13}$$

the variance of the estimator is,

$$\text{var}(\mu) = \frac{\text{var}[f(x)]}{n},$$

The MC estimator does not work accurately for large variance. Statistical errors in the confidence interval can be too large to believe that the interval is valid. For

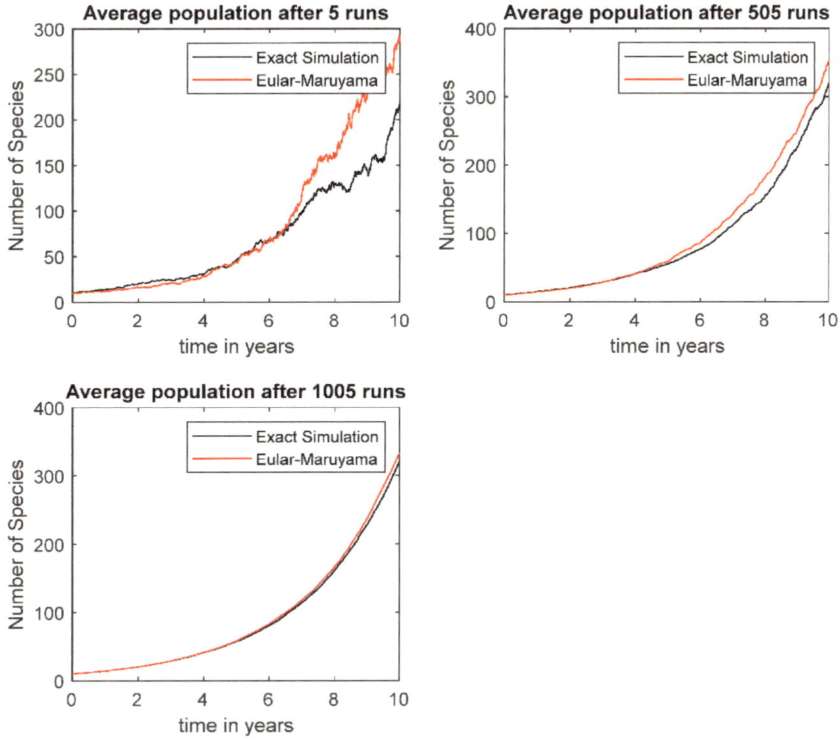

Figure 3.
Simulation results for the exact and Euler-Maruyama approximation after a number of runs.

variance reduction, standard deviation of the estimator equals to $\frac{\sigma_t}{\sqrt{n}}$. For small confidence interval, we can reduce the variance. For example, $\text{var}[f(x)] = 1$, and a smaller confidence interval is required, say 10^{-2};

Thus,

$$\text{var}[f(x)] = 10^{-4}(n) \tag{14}$$

Thus, we need $n = 10^4$ sample points. Numerous techniques based on the control variate, importance sampling, stratified sampling and anthetic variates are available in literature. Zhang and Karniadakis [29] used different reduction methods to avoid generating a large number of sample points. The convergence rate of the basic MC estimator for a one-dimensional integral is $O\left(\frac{1}{\sqrt{n}}\right)$. While this convergence is free from stochastic dimensions, it is slow. The positive side of this independence confirms the applicability of MC methods in problems with high-stochastic dimensions.

2.4 Quasi monte carlo (QMC) methods

Computing mathematical models using MC methods is expensive due to the requirement of a large number of sample points. One solution is to use variance reduction techniques. An error in these approaches is proportional to $\frac{1}{n}$ as compared to

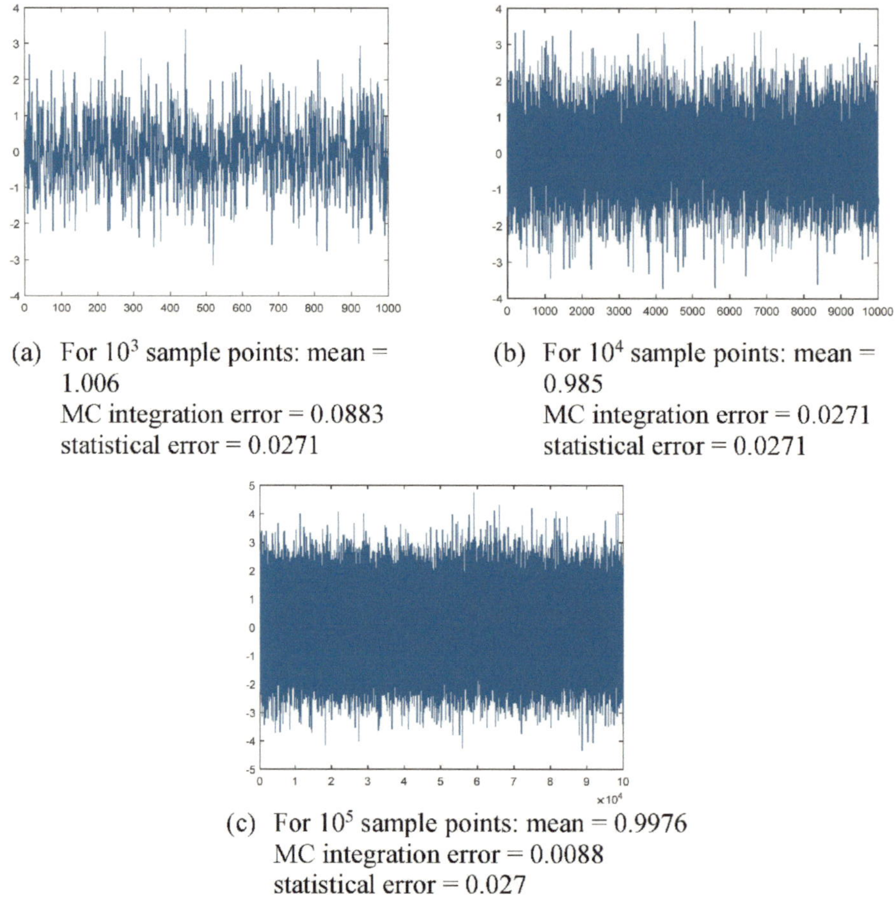

Figure 4.
Generation of a Gaussian random variate (GRV) for different sample sizes and a comparison with MC integration error and the mean.

$\frac{1}{\sqrt{n}}$ as in the MC methods. A detailed study of these techniques has been given in [30]. A few low-discrepancy sequences are shown in **Figure 5**.

2.5 Stochastic spectral methods

Stochastic spectral methods use spectral representations to efficiently capture the inherent randomness and complexity in systems governed by SDEs and SPDEs. Some of the spectral methods are:

2.5.1 Karhunen-Loève expansion

The Karhunen-Loève expansion is particularly useful for problems with random spatial variations. It uses eigenfunctions to represent stochastic processes and helps in dimensionality reduction. In environmental modelling, this method is being applied to simulate groundwater flow in heterogeneous aquifers, where the spatial variability of hydraulic conductivity is considered as a stochastic field.

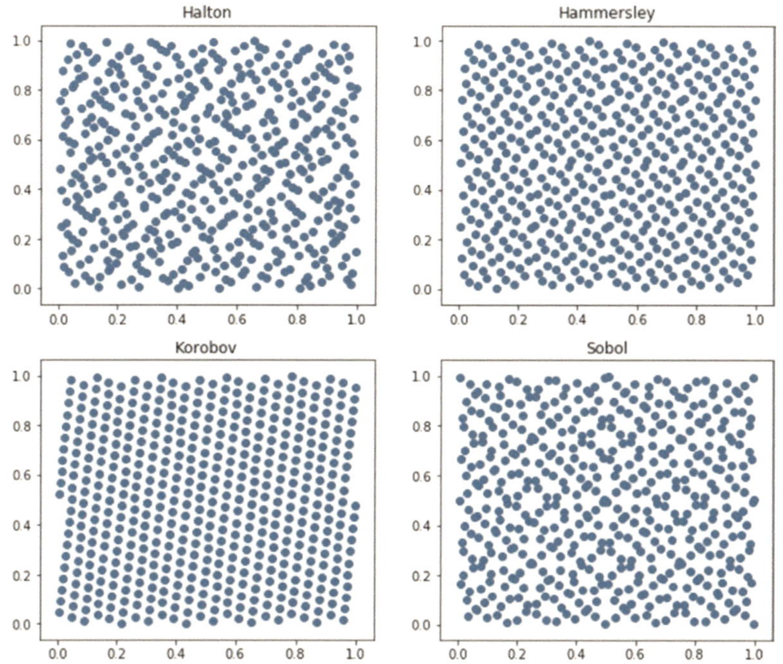

Figure 5.
Quasi-random sequences. Quasi-random sequences are used in variance reduction techniques. The word 'quasi' signifies that the values of a low-discrepancy sequence is neither random nor pseudo-random, but share some of the characteristics of both of the random variables. This low-discrepancy property is an added benefit of QMC methods. Using a control variate helps to reduce MC variance.

2.5.2 Polynomial chaos expansion (PCE) methods

The basic idea of PCE methods is to approximate the solution as a sum of polynomial functions that are orthogonal, with respect to the probability distribution of the random inputs [25]. The coefficients of the expansion are determined by solving a system of equations derived from the original problem, typically using techniques such as Galerkin projection or least-squares regression. A schematic for working process of PCE methods is shown in **Figure 6**.

The resulting PCE can then be used to estimate the solution at any point in the domain, including those with stochastic inputs. One advantage of PCE methods is in their efficiency to propagate uncertainty through the system, allowing for the efficient estimation of statistics such as mean and variance [31]. PCE methods can also be used to efficiently quantify the effect of input uncertainties on the system response, such as in sensitivity analysis. PCE methods can be used in both intrusive and non-intrusive implementations. While intrusive PCE methods modify the original deterministic solver to handle the stochastic inputs directly, non-intrusive PCE methods use a surrogate model to represent the solution in terms of the PCE.

2.5.3 Generalised polynomial chaos expansion (gPCE) methods

Wiener PCE methods use Hermite polynomials and express the solution of SDE/SPDE in terms of these smoothed orthogonal polynomials. Although these homogeneous Wiener polynomial chaos is quite useful and satisfies certain universal

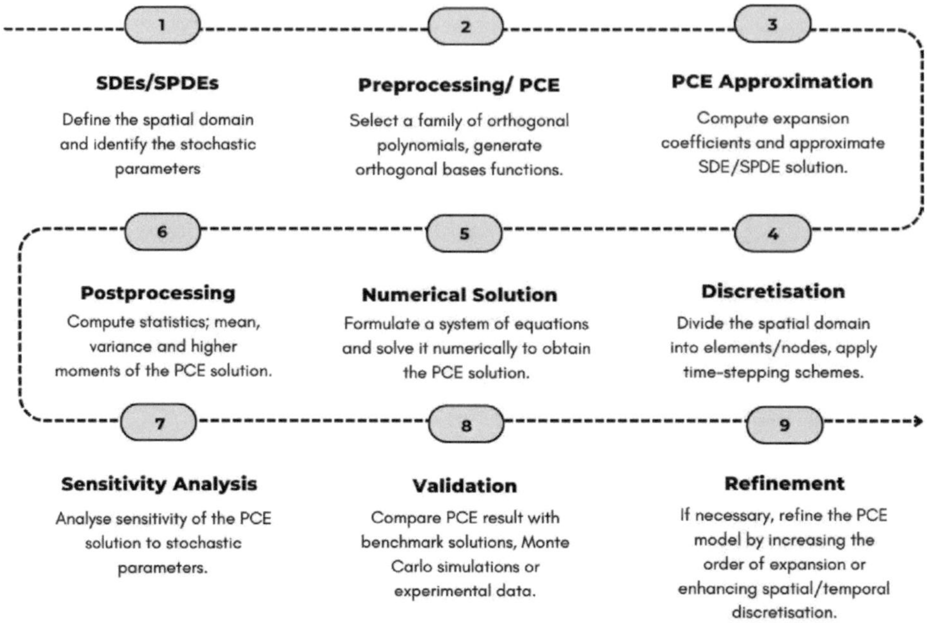

Figure 6.
Diagrammatic representation for working process of polynomial chaos expansion methods for SDEs and SPDEs.

polynomial characteristics, but these polynomials are more useful for Gaussian distributed random input parameters. For non-Gaussian random variables, Hermite polynomial-based chaos expansion is slow. Askey scheme for orthogonal polynomials is given in [32]. It involves mapping the stochastic solution onto a set of polynomials that are orthogonal to the probability distribution of the input parameter.

Thus, instead of finding the random solution of the SDE/SPDE, the problem is reduced to evaluating the expansion coefficients. The basis functions are selected based on the distribution of random parameters. The correspondence between the basis functions and their known distributions is available in the literature [32] and is summarised in **Figure 7**.

3. Applications of PCE method

3.1 Development of a meta model for geometric brownian motion (gBM)

A sequence of random variables $W = \{W(t)\}_{t\geq 0} \in R$ in a time domain on a probability space $(\Omega, F, \wp)$ is known as a standard Brownian motion.

For a fixed $T \in (0, \infty)$ and a filteration, let F_t^W be the sigma-algebra generated by these random variables for $0 \leq t \leq T$. Using Cameron-Martin basis [33] for a countable multi-index set,

$$J = \left\{ \left(\alpha_i^j, i, j \geq 1 \right), \alpha_i^j \in \{0, 1, \ldots \ldots\} \right\} \text{with } \alpha! = \prod_{i,j} \alpha_i^j! \tag{15}$$

here i and j are the number of Gaussian random variables and the Wiener processes, respectively.

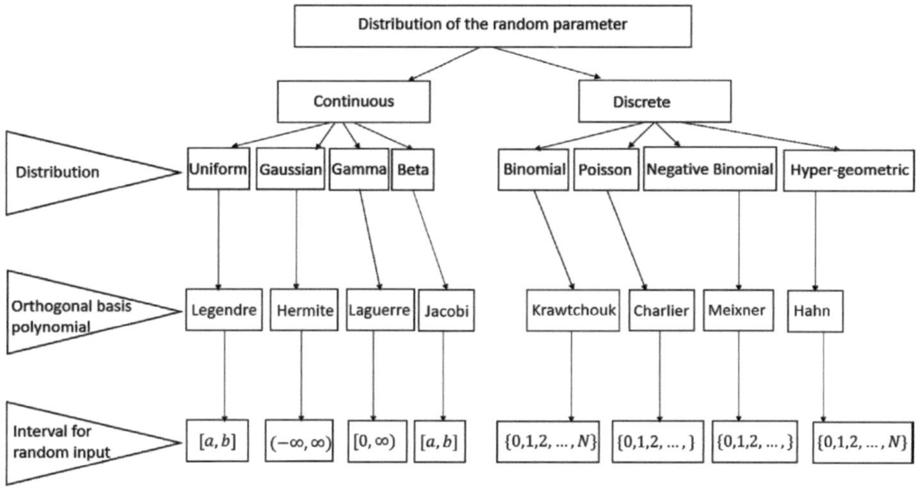

Figure 7.
Equivalence between distributions, basis polynomials and related input random variables.

According to the Cameron-Martin theorem, there exists an orthonormal basis $m_i, i \geq 1$ of square integrable space $L^2([0, T])$, such that a set of independent random variables ξ_{ij} is defined as,

$$\xi_{ij} = \int_0^T m_i(\tau) dW_\tau^j \tag{16}$$

Now, the set of random variables known as Cameron-Martin basis are defined as,

$$T = \{\xi_\alpha, \alpha \in J\} \tag{17}$$

A space generated by this basis is the Wiener chaos space w and a random variable $f(\omega) \in w$, or some realisation ω is written in terms of Wiener chaos expansion as,

$$f = \sum_{\alpha \in J} a_\alpha \xi_\alpha \tag{18}$$

With the above setup defined, we now develop a meta model for geometric Brownian motion given by the solution of SDE, defined in Eq. (19).

Again,

$$dY_t = \mu(t, Y_t)dt + \sigma(t, Y_t)dW_t \tag{19}$$

gBM is the solution of Eq. (19).

Applying the Euler-Maruyama method and using independent increments of the Wiener process and for each increment $m = \{0, 1, \dots \dots, M - 1\}$, we have,

$$Y_{t_{m+1}} = Y_{t_m} \tag{20}$$

for $0 = t_1 < t_2 < \dots \dots < t_{M-1} < t_M = T$ and $\Delta t_m = t_{m+1} - t_m$.

For meta modelling, we input this multivariate function in the form,

$$Y_{t_m} = M(Y) \tag{21}$$
$$Y = (\xi_1, \xi_2, \dots \dots \dots, \xi_{m-1}), \xi_i \, N(0,1)$$

and, for the Euler-Maruyama scheme,

$$W(t) \, N(0,t), t \in [0,T]$$

This model automatically chooses Hermite polynomials as the basis functions since the random variables are Gaussian. More conveniently, Eq. (20) is written as

$$Y_t = Y_{t_m} = Y_0 \prod_{m=0}^{M-1} \left(1 + \mu(t_m, Y_{t_m})\Delta t_m + \sigma(t_m, Y_{t_m})\sqrt{\Delta t_m \xi_m}\right) \tag{22}$$

Using a suitable transformation, gBM converts Eq. (19) into the form of Eq. (21). Using the Itô's lemma and defining $X_t = ln\, S_t$, we get

$$dX_t = \left(\mu - \frac{1}{2}\right)\sigma^2 dt + \sigma(t, X_t)dW_t \tag{23}$$

Now, for any computed value of X_t, we have,

$$S_t = e^{X_t} \tag{24}$$

and this output is of the form

$$S_T = M(Y, \odot) \tag{25}$$

where $\odot$ is the transformation.

Using a UQLab framework [34] and feeding input Y as gBM, we develop a meta model. **Figure 8** represents the PCE coefficients spectrum based on different methods. The hyperbolic truncation scheme is used with q-norm in the range $0 \leq q - \text{norm} \leq 1$. The Sobol sequence sampling strategy is used for all methods. The graphs show that the number of PCE coefficients can be reduced using a suitable evaluation method and the sparse sampling technique. The output of the model is summarised in **Table 1**.

This model is also validated with a set of 1E4 samples. The results for the validation error using different approaches are provided in **Table 2**.

3.2 Stochastic contaminant transport

The stochastic advection dispersion equation for contaminant transport [35] is given below:

$$\frac{\partial C}{\partial t} + \mu.\nabla C = \nabla.[\kappa(x;\omega)\nabla C] + F(t,x;\omega), (t,x;\omega) \in R^+ \times D \times \Omega \tag{26}$$

with the initial condition $C(0,x;\omega) = C_0(x;\omega)$, where C is the concentration of the contaminantwithin the bounded domain $D \in R^d (d = 1, 2)$, R^+ is the time domain $(0, T]$ and Ω is the sample space. $F(t,x;\omega)$ is the random forcing term. We consider $F(t,x;\omega) = \sigma \frac{\partial W}{\partial t}$, where σ is the standard deviation of the stochastic term, and $\frac{\partial W}{\partial t}$

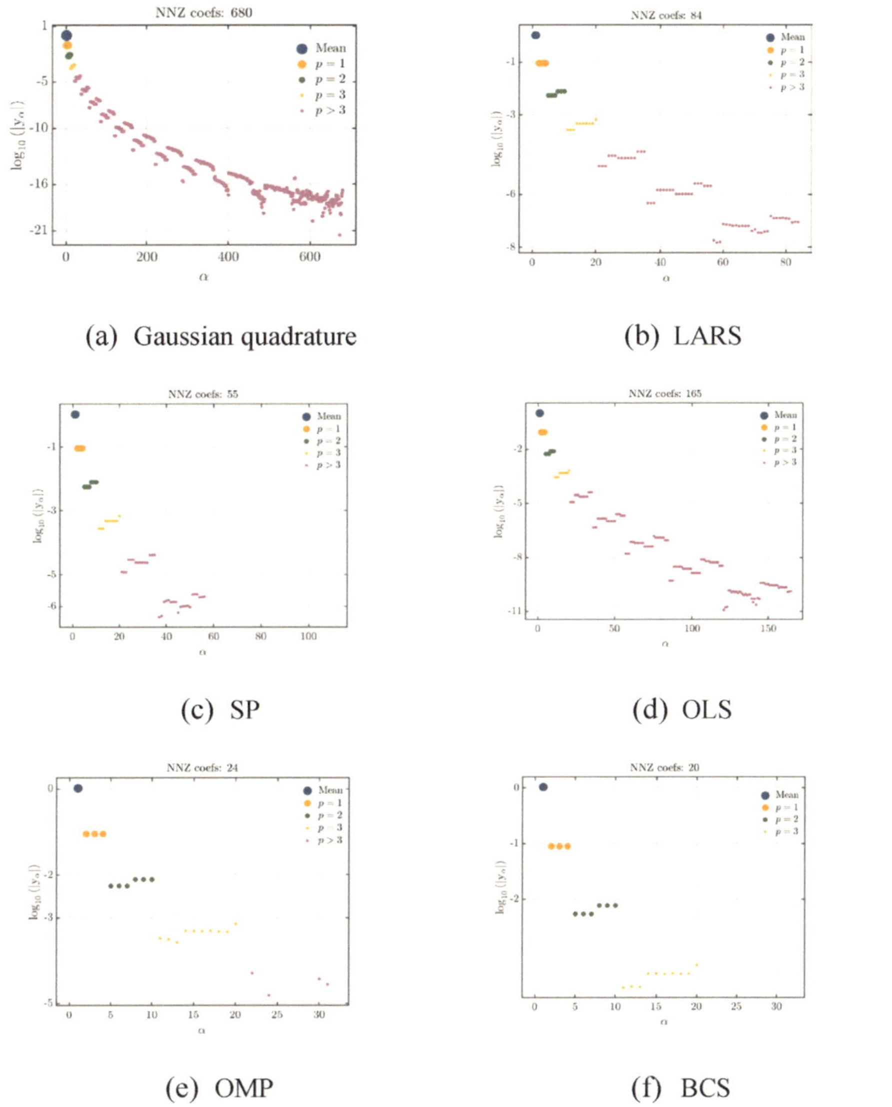

Figure 8.
Spectrum of the PCE coefficients for the meta modelling of geometric Brownian motion using (a) Gaussian quadrature, (b) least angle regression, (c) subspace, (d) ordinary least-squares regression, (e) orthogonal matching pursuit and (f) Bayesian compressive sensing. NNZ represents the number of nonzero coefficients in the chaos expansion.

represents the time derivative of a Wiener process which introduces stochastic variability. μ is the mean advection velocity representing the bulk flow of the fluid; $\kappa(x;\omega)$ is the dispersion coefficient characterising the rate of diffusion and $C_0(x;\omega)$ is the initial condition. $\kappa(x;\omega)$ and $C_0(x;\omega)$ are both real valued functions in D and Ω for all random inputs.

In gPCE, random numbers are chosen based on the probability distribution of input variables and corresponding orthogonal basis functions. Choosing $\{\varphi_n\}, n = 0, 1, \ldots \ldots$ as orthogonal polynomials for random variable ξ satisfies the orthogonality condition given. Applying gPCE to Eq. (26), we have

Parameters	Method					
	Gaussian Quadrature	BCS	LARS	SP	OMP	OLS
Number of input variables	3	3	3	3	3	3
Maximal degree	14	8	6	9	5	5
q-norm	1	1	1	0.75	0.75	0.75
Size of full basis	680	165	84	111	32	32
Size of sparse basis	680	165	84	55	24	20
Full model evaluations	3375	500	150	150	150	150
Quadrature error	2.24E-29	2.54E-19	1.04E-13	8.13E-12	3.52E-07	1.43E-07
Mean value	1.0305	1.0305	1.0305	1.0305	1.0305	1.0305
Standard deviation	0.1554	0.1554	0.1554	0.1554	0.1554	0.1554
Coefficient of variation	15.09%	15.09%	15.09%	15.09%	15.09%	15.09%

Table 1.
Meta model's input and output parameters.

Validation parameter	Gaussian Quadrature	BCS	LARS	SP	OMP	OLS
Relative error	4.59E-28	1.56E-18	1.27E-13	1.10E-06	7.54E-11	6.92E-07

Table 2.
Relative error for model validation.

$$\kappa(x;\omega) = \sum_{i=0}^{N} \kappa_i(x,t)\varphi_i(\xi) \tag{27}$$

$$C(t,x;\xi) = \sum_{j=0}^{N} C_j(t,x)\varphi_j(\xi) \tag{28}$$

and

$$F(t,x;\xi) = \sum_{j=0}^{N} F_j(t,x)\varphi_j(\xi) \tag{29}$$

The value of N (the total number of expansion terms) is based on the order of polynomials (n) used and the dimension of the random space (p). It is given by,

$$N+1 = \frac{(n+p)!}{n!p!} \tag{30}$$

Substituting Eq. (27) to Eq. (29) in the governing Eq. (26), we get

$$\begin{aligned}\sum_{j=0}^{N} \frac{\partial C_j(t,x)}{\partial t}\varphi_j(\xi) + \mu\sum_{j=0}^{N} \nabla C_j(t,x)\varphi_j(\xi) \\ = \nabla.\left[\sum_{i=0}^{N}\kappa_i(x)\varphi_i(\xi)\nabla\left(\sum_{j=0}^{N} C_j(t,x)\varphi_j(\xi)\right)\right] + \sum_{j=0}^{N} F_j(t,x)\varphi_j(\xi)\end{aligned} \tag{31}$$

Rewriting the above equation,

$$\sum_{j=0}^{N} \frac{\partial C_j(t,x)}{\partial t} \varphi_j + \mu \sum_{j=0}^{N} \nabla C_j(t,x)\varphi_j = \left[\sum_{i=0}^{N}\sum_{j=0}^{N} \nabla.\left[\kappa_i(x)\nabla C_j(t,x)\right]\right]\varphi_i\varphi_j + \sum_{j=0}^{N} F_j(t,x)\varphi_j \tag{32}$$

Next, applying the Galerkin projection onto each polynomial basis ensures that the induced error is orthogonal to functional space spanned by truncated finite dimensional basis functions φ_n. Projecting φ_m for each $m = \{0, 1, \ldots \ldots, N\}$ and applying the orthogonality conditions, we get

$$\frac{\partial C_m}{\partial t} + \mu \nabla C_m = \sum_{i=0}^{N}\sum_{j=0}^{N} \nabla.\left[\kappa_i(x)\nabla C_j\right]E_{imj} + F_m \tag{33}$$

where $E_{imj} = <\varphi_i\varphi_k\varphi_j>$ and is calculated as $E_{imj} = \frac{<\varphi_i\varphi_j, \varphi_m>}{<\varphi_m, \varphi_m>}$ by setting up the definition of φ_n. The above equation can be written as,

$$\frac{\partial C_m}{\partial t} + \mu \nabla C_m = \sum_{j=0}^{N} \nabla.\left[a_{mj}(x)\nabla C_j\right] + F_m \tag{34}$$

or

$$\frac{\partial C_m}{\partial t} = \sum_{j=0}^{N} \left[a_{mj}(x)\nabla^2 C_j + b_{mj}(x).\nabla C_m\right] - \mu.\nabla C_m + F_m, m = 0, 1, \ldots \ldots \ldots, N \tag{35}$$

where,

$$a_{mj}(x) = \sum_{i=0}^{N} \kappa_i(x)E_{imj} \tag{36}$$

and,

$$b_{mj}(x) = \sum_{i=0}^{N} \nabla\kappa_i(x)E_{imj} = \nabla a_{mj}(x) \tag{37}$$

Equation (35) is a set of (N + 1) coupled deterministic equations with an initial condition,

$$C(0, x; \omega) = \sum_{j=0}^{N} C_j(0, x)\varphi(\xi) = C_0(x; \omega)$$

where the coefficients $C_j(0, x) = C_0(x)$ in the expansion are deterministic and we can obtain the initial and boundary conditions for each expanded (Eq. (26)). For numerical simulation, we considered a one-dimensional case and the equation,

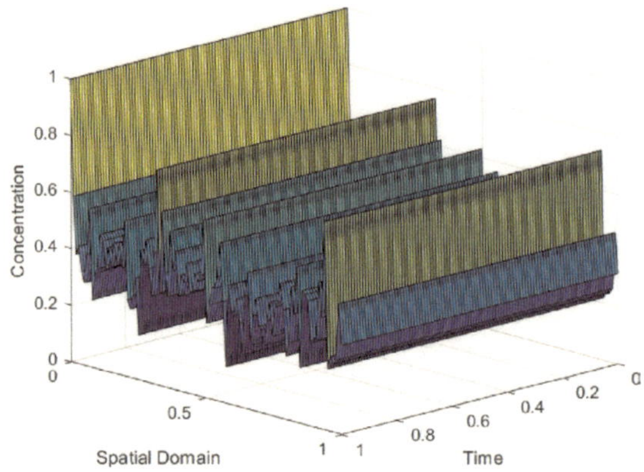

Figure 9.
Simulation of concentration using MC method.

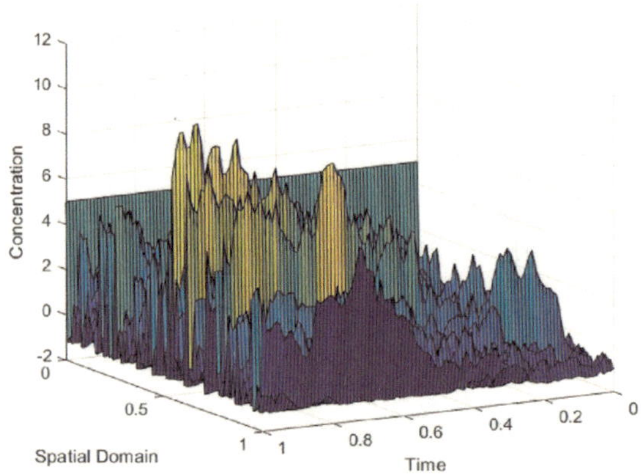

Figure 10.
Simulation of concentration using PCE method.

$$\frac{dC(t,x;\omega)}{dt}+\mu\frac{dC(t,x;\omega)}{dx}=\frac{d}{dt}\left[\kappa(x;\omega)\frac{dC(t,x;\omega)}{dx}\right]+F(t,x;\omega) \tag{38}$$

with the deterministic initial condition $C(0,x;\omega)=1$ and boundary conditions $x\in[0,1)$. Without a loss of generality, randomness in transport is characterised by considering the random forcing term as $F(t,x;\omega)=\sigma\frac{\partial W}{\partial t}$ with the Wiener process term and $\kappa(x;\omega)$ is characterised with $\kappa(x;\omega)=\xi$. While applying MC simulation, 10^5 random numbers are generated. For polynomial chaos expansion, the randomness in σ is captured through the polynomial basis functions and the Wiener process. **Figures 9** and **10** show the concentration profile using the MC method with 10^4 samples and the PCE method of polynomial order 3 for $C(0,x;\omega)=1$.

Time domain R^+ is (0,1] with time step $\Delta t=0.01$, $C(0,x;\omega)=1$, space domain $x\in[0,1)$ and number of random samples 10000. Hermite polynomials are of order 3 and maximum order of polynomial chaos expansion is 5. The mean and standard deviation of the Gaussian distribution are 0 and 1, respectively. **Figure 10** shows that PCE method is more efficient in capturing the randomness in $C(t,x;\omega)$ compared to MC simulation.

3.2.1 Effect of boundary conditions on solution

The choice of boundary conditions plays an important role in determining the appropriate order of polynomial chaos expansion (PCE) for computer simulations.

Graphs for the Gaussian initial condition $C(0,x;\omega) = \frac{1}{\sigma_G\sqrt{2\pi}} exp\left((x-\mu)^2/2\sigma_G\right)$ for different order of Hermite polynomials are shown in **Figure 11(a)–(d)**. Boundary conditions determine the behaviour of a random process at the boundaries of the domain of interest. For example, imposing periodic or reflection boundary conditions can have a significant impact on the statistical properties of the random field. In cases where the boundary conditions introduce additional sources of randomness or nonlinearity, higher-order PCE may be needed to accurately capture these effects.

In contrast, simpler boundary conditions, such as Dirichlet or Neumann conditions, can lead to smoother responses, allowing lower-order PCEs to provide accurate approximations. Therefore, choosing the appropriate PCE order is closely related to understanding the boundary conditions and their influence on the stochastic

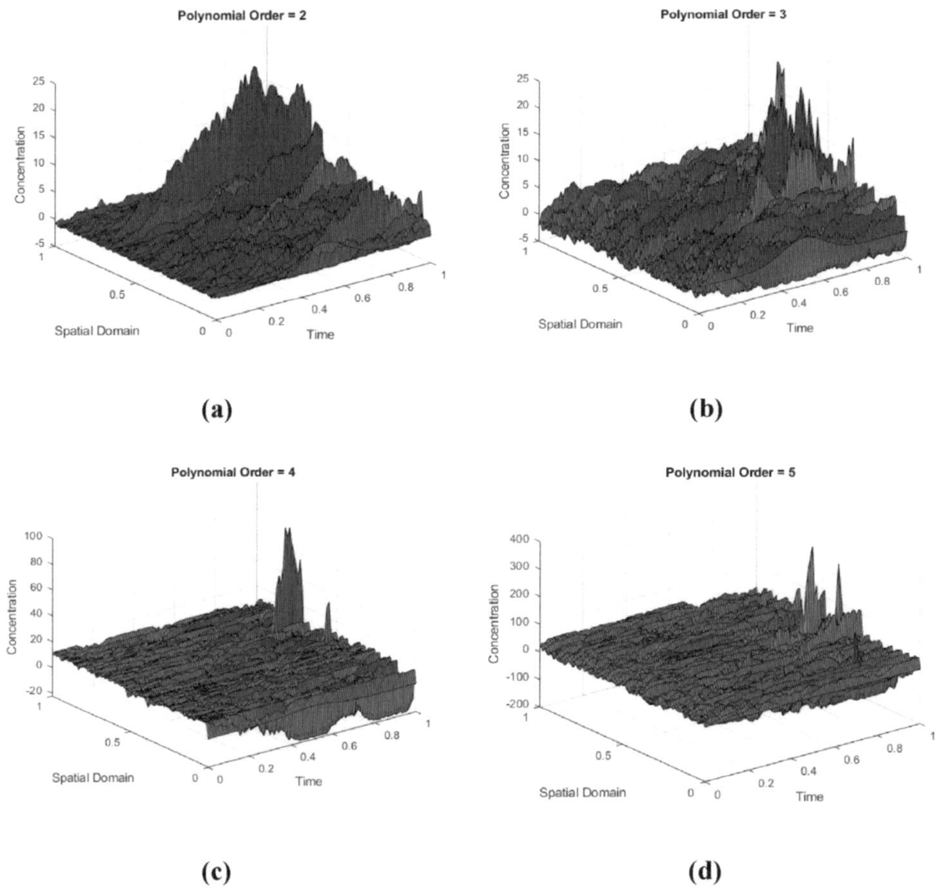

Figure 11.
Simulation of concentration C using the PCE method with the Gaussian initial condition and various order of polynomial chaos expansion. Time domain R^+ is $(0,1]$, time step $\Delta t = 0.01$, Hermite polynomials of order 3. The mean and standard deviation of Gaussian distribution is $\mu = 0.5$ and standard deviation $\sigma_G = 0.1$.

behaviour of the system, thereby ensuring that the computational model faithfully represents the physical reality of the problem. The graphs in **Figure 11** show that lower-order polynomials are capable of capturing the Gaussian distribution more accurately than higher-order polynomials.

4. Wick product

The Wick product was initially used as a renormalisation process. T. Hida and N. Ikeda [36] proposed the Wick product for stochastic analysis. Later on, this concept was expanded [37] to include white noise in the Wick products of stochastic distributions and applied Gaussian Wick calculus. A detailed and clear overview of the Wick product and Wick renormalisation is explained in [38].

4.1 Motivation for Wick multiplication (renormalisation point of view)

White noise $W(t)$ can be represented as an element of Hida dual space $\left(L^2\right)^{-1}$ as,

$$W(t) = \lim_{\Delta t \to 0} \frac{1}{\Delta t} \int_t^{t+\Delta t} dB_u \tag{39}$$

$$\int_R \delta_t dB_u \tag{40}$$

where δ_t is the Dirac measure of t that belongs to the dual Sobolov space. Using Itô's formula,

$$(W_t)^2 \approx \left(\frac{1}{\Delta t} \int_t^{t+\Delta t} dB_u \right)^2 \tag{41}$$

$$\frac{2}{(\Delta t)^2} \int_t^{t+\Delta t} \left(\left(\int_t^v dB_u \right) dB_v + \frac{1}{\Delta t} \right) \tag{42}$$

The additive re-normalisation of the above term is,

$$\frac{2}{(\Delta t)^2} \int_t^{t+\Delta t} \left(\left(\int_t^v dB_u \right) dB_v \right)$$

This motivates the definition of,

$$\int \delta_t dB \Diamond \delta_t dB = \int \delta_t \otimes \delta_t dB^{\otimes 2} \tag{43}$$

Here, $\otimes$ is the symmetrized tensor product. The generalised structure of the solutions is transferred to the stochastic component, and the SPDEs are interpreted in the typical strong sense with respect to the time parameter t and the spatial parameter $x \in R^d$. An important aspect of using the Wick product is the ability to define a unique white noise process as a mathematical object, such as the time derivative of the Brownian motion. SDEs can therefore be solved as true differential equations, which is how they are typically understood, rather than just as integral equations.

The Wick product is used in the context of white noise analysis to solve the issue of pointwise multiplication of generalised functions. It is specified in the test and generalised stochastic function of Kondratiev spaces. It is also useful to replace all products with Wick products and all nonlinear functions with their Wick equivalents in order to solve the multiplicative noise and nonlinear equations because the Wick product is a product in the algebraic sense. Although the Wick product performs very smoothly as a mathematical object, care should be taken when using it.

4.2 The Wick product in stochastic analysis

If $(S)^*$ is a Hida distribution space and $f(\omega)$ and $g(\omega)$ are two elements of $(S)^*$ defined as,

$$f(\omega) = \sum\nolimits_{\alpha} a_\alpha H_\alpha \tag{44}$$

and

$$g(\omega) = \sum\nolimits_{\beta} b_\beta H_\beta \tag{45}$$

where α and β are the multi-indices, then the Wick product of $f(\omega)$ and $g(w)$ is defined as,

$$f(\omega) \Diamond g(\omega) = \sum\nolimits_{\alpha,\beta} a_\alpha b_\beta H_{\alpha+\beta} \tag{46}$$

Also, $f(\omega) \Diamond g(\omega) \in (S)^*$. A detailed explanation and application of the Wick product can be found in the extant literature [39].

4.3 Properties of the Wick product

The following are some of the properties of the Wick product used in stochastic analysis:

$$W^{\Diamond 2}(t) = W^2(t) - t, t \geq 0$$

For a random process X, $E\left[\exp^{\Diamond}(X)\right] = \exp\mathrm{E}[\mathrm{X}]$

$$E\left[X \Diamond Y\right] = E[X]E[Y]$$

Chain rule: for an analytic function $g : C \to C, i.e., g(R) \subseteq R$,

$$\frac{d}{dt}\left[g^{\Diamond}(x(t))\right] = (g')^{\Diamond}\left(x(t)^{\Diamond} x'(t)\right) \tag{47}$$

also,

$$\left(\int_R g(x)dW(x)\right)\Diamond\left(\int_R h(y)dW(y)\right) = \left(\int_R g(x)dW(x)\right)\left(\int_R h(y)dW(y)\right) - \int_R g(t)h(t)dt, \tag{48}$$

$\forall g, h \in L^2(R)$ and are deterministic.

In general, the nth Wick exponential of ϑ is defined as

$$\vartheta^{\Diamond n} = ||f||^n H_n\left(\frac{\vartheta}{||f||}\right) \tag{49}$$

where $\vartheta = \int_R f dW$ and hence,

$$W^{\Diamond n}(t) = t^{\frac{n}{2}} H_n\left(\frac{W(t)}{t^{\frac{1}{2}}}\right), n = 0, 1, 2, \dots \dots \dots \tag{50}$$

$$\exp^{\Diamond}\left\{\int_R f dW\right\} = \exp\left\{\int_R f dW - \frac{1}{2}||f||^2\right\}, f \in L^2(R) \tag{51}$$

and,

$$\exp^{\Diamond} W(t) = \exp\left\{W(t) - \frac{1}{2}t\right\}, \forall t \geq 0 \tag{52}$$

Equations (49) to (52) are very important from computational point of view. These equations help us to understand the basic building block of smoothed white noise process and need of Wick product.

4.4 Computation of wick product

Let us consider (Eq. 52) and compute the Wick product and Wick exponential. First, we will compute least-square norm of a deterministic function $f(t)$, that is, $f \in L^2(R)$ given by a second-degree polynomial

$$f(t) = a + bt + ct^2 \tag{53}$$

L^2-norm of $f(t)$ is

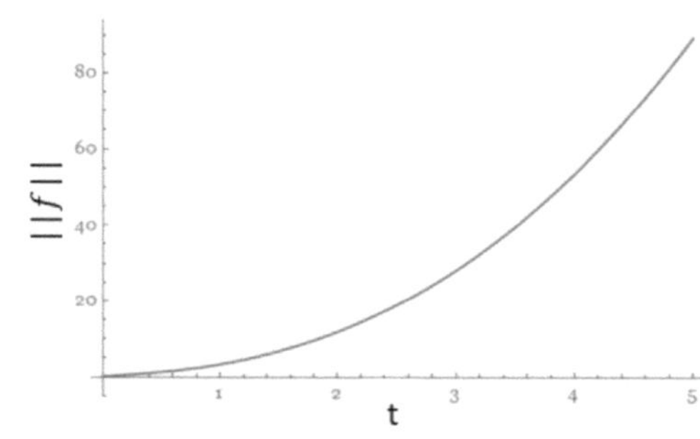

Figure 12.
Least square norm of a second-degree polynomial function.

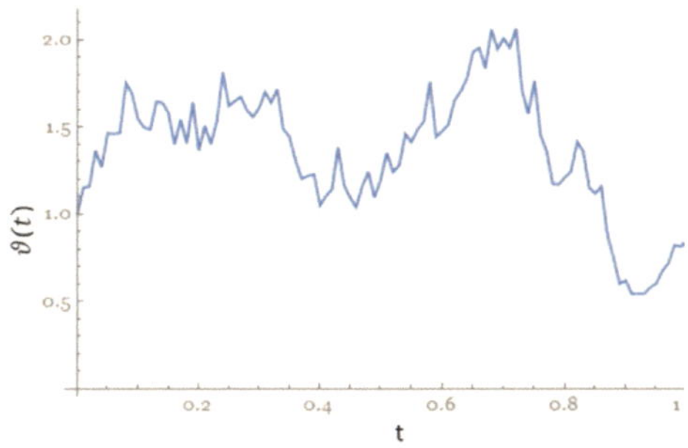

Figure 13.
Realisation of an Itô process.

$$\|f\| = \left(\int_0^t (f(t))^2 dt \right)^{1/2} \tag{54}$$

When $f(t) = 1$, $\|f\| = \left(\int_0^t dt \right)^{1/2} = \sqrt{t}$, when $a = 0, b = 1, c = 0$; $\|f\| = \left(\int_0^t (f(t))^2 dt \right)^{1/2} = \sqrt{\frac{t^3}{3}}$ and when $a = 1, b = 2, c = 0.001$; $\|f\| = \sqrt{2 \times 10^{-7} t^5 + 0.001t^4 + 1.334t^3 + 2t^2 + t}$ and is plotted in **Figure 12**.

Now, let $\vartheta(t) = \int_R f(t) dW$; this is an Itô process, and a realisation of ϑ for the interval $[0, 1]$ is shown in **Figure 13**.

Let $\Psi(t) = \frac{\vartheta}{\|f\|}$ so that using (Eq. 50), we have,

$$\Psi(t) = \frac{\int_0^t f(t) dW(t)}{\|f\|} \tag{55}$$

As we have already computed the numerator $\int_0^t f(t) dW(t)$ and denominator $\|f\|$, we can compute $\Psi(t)$, and then, incorporating this $\Psi(t)$ into (Eq. 50) we can compute $\vartheta^{\Diamond n}$ as

$$\vartheta^{\Diamond n} = \|f\|^n H_n(\Psi(t))$$

Further, using (Eq. 52), we have

$$\exp^{\Diamond}\left\{ \int_R f(t) dW(t) \right\} = \exp^{\Diamond}\{\vartheta(t)\} = \exp\left\{ \vartheta(t) - \frac{1}{2}\|f\|^2 \right\} \tag{56}$$

also, if $f(t) = 1$,

$$\exp^{\Diamond}\left\{ \int_R f(t) dW(t) \right\} = \exp^{\Diamond}\{W(t)\} = \exp\left\{ W(t) - \frac{1}{2}t \right\} \tag{57}$$

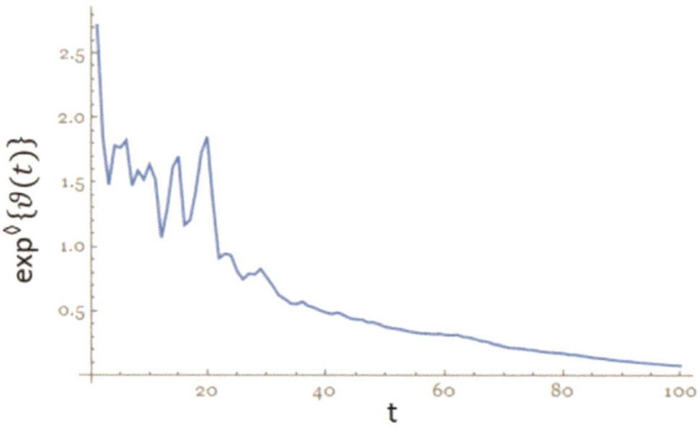

Figure 14.
Wick exponential of an Itô process $\vartheta(t) = \int_R f(t)dW(t), f(t) = a + bt + ct^2$ *with* $a = 1, b = 2,$ *and* $c = 0.001$.

for $f(t) = 1 + 2t + 0.001t^2$, the graph for Wick exponential (Eq. 55) is plotted in **Figure 14**.

5. Discussion and conclusion

Numerical approaches offer a systematic and effective way to simulate the dynamic behaviour of complex systems, allowing academics and practitioners to obtain insights into their complex dynamics. Numerical methods are vital tools for modelling and analysis of these systems, where the interaction of many elements creates stochasticity. Finite difference/element methods, Monte Carlo simulations and spectral methods are all adaptable approaches to dealing with SDEs and SPDEs. These methods assist the investigation of complex events in a variety of domains by providing computationally feasible solutions.

The ability of numerical approaches to deal with high-dimensional problems and spatial changes is very important when studying complex systems. For example, Stochastic Collocation and Galerkin approaches enable researchers to address uncertainty in several dimensions and spatial domains. This adaptability is essential for reflecting the complexities of real-world systems where uncertainties emerge in a variety of ways. Furthermore, as computer capabilities improve, numerical approaches play an increasingly important role in enhancing our knowledge of complex systems. These give a practical way to investigate and simulate stochastic systems, allowing for the exploration of problems that would otherwise be intractable using analytical methods alone.

The toolbox of stochastic spectral approaches is extensive and diverse. While stochastic spectral approaches provide an impressive set of tools for dealing with uncertainty, these approaches are not without limitations. One of the most significant issues is the computational intensity, particularly when dealing with high-dimensional situations. Monte Carlo simulations, for example, may require a large number of samples to reach convergence, making them computationally expensive. Similarly, polynomial chaos expansions have difficulties in adequately capturing highly nonlinear dynamics, frequently demanding an increased number of terms or advanced techniques to accurately approximate the solution. As a result, these methods may struggle to give real-time solutions for complicated systems, limiting their application in situations requiring quick decision-making.

Another drawback comes from the inbuilt assumptions of the modelling process. Numerical methods often assume certain statistical properties of the underlying uncertainties, such as square integrable spaces, specific probability distributions or correlations. These assumptions, while simplifying the problem, may not always align with the true nature of the system, leading to model inaccuracies. Also, the curse of dimensionality remains a serious issue, especially in high-dimensional systems, where it can exponentially increase the computational cost. As a result, when applied to real-world scenarios with large parameter spaces, numerical methods may find practical challenges. To address these constraints, continued research into more efficient algorithms, adaptive strategies and novel methodologies to improve the adaptability and dependability of numerical approaches in dealing with complex, high-dimensional and nonlinear systems is required.

Author details

Parul Tiwari[1,2], Don Kulasiri[2]* and Sandhya Samarasinghe[2]

1 School of Engineering, Computer and Mathematical Sciences, Auckland University of Technology, Auckland, New Zealand

2 Center for Advanced Computational Solutions, Lincoln University, Christchurch, New Zealand

*Address all correspondence to: don.kulasiri@lincoln.ac.nz

References

[1] Nikulina TN, Zhirnova IS, Stupina AA, Zhirnov AA. Mathematical modeling of economic processes in complex systems (on the example of Krasnoyarsk municipality). Journal of Physics Conference Series. 2019;**1353**: 12118

[2] Zhang D, Lu L, Guo L, Karniadakis GE. Quantifying Total Uncertainty in Physics-Informed Neural Networks for Solving Forward and Inverse Stochastic Problems. 2018. Available from: http://arxiv.org/abs/1809.08327

[3] Lunz D, Batt G, Ruess J, Bonnans JF. Beyond the chemical master equation: Stochastic chemical kinetics coupled with auxiliary processes. PLoS Computational Biology [Internet]. 2021; 7(7):e1009214. DOI: 10.1371/journal.pcbi.1009214

[4] Kosarwal R, Kulasiri D, Samarasinghe S. Novel domain expansion methods to improve the computational efficiency of the chemical master equation solution for large biological networks. BMC Bioinformatics [Internet]. 2020;**21**(1):1-42. Available from: https://bmcbioinformatics.biomedcentral.com/articles/10.1186/s12859-020-03668-2

[5] Tryoen J, Le MO, Ndjinga M, Ern A. Intrusive Galerkin methods with upwinding for uncertain nonlinear hyperbolic systems. Journal of Computational Physics [Internet]. 2010; **229**(18):6485-6511. Available from: https://research.kaust.edu.sa/en/publications/intrusive-galerkin-methods-with-upwinding-for-uncertain-nonlinear

[6] Ghanem R, Owhadi H, Higdon D. Handbook of uncertainty quantification. Handbook of Uncertainty Quantification. 2017;**16**:1-2053

[7] Heston SL. A closed-form solution for options with stochastic volatility with applications to bond and currency options. The Review of Financial Studies [Internet]. 1993;**6**(2):327-343. Available from: https://academic.oup.com/rfs/article/6/2/327/1574747

[8] Zhang JE, Shu J. Pricing S&P 500 index options with Heston's model. In: IEEE/IAFE Conference on Computational Intelligence for Financial Engineering, Proceedings (CIFEr). 2003; 2003-January:85–92

[9] Faria G, Correia-da-Silva J. A closed-form solution for options with ambiguity about stochastic volatility. Review of Derivatives Research. 2014;**17**(2): 125-159. Available from: https://link.springer.com/article/10.1007/s11147-014-9097-9

[10] Yoshioka H, Unami K, Kawachi T. Stochastic process model for solute transport and the associated transport equation. Applied Mathematical Modelling. 2012;**36**(4):1796-1805. DOI: 10.1016/j.apm.2011.09.011

[11] Ohsumi A. An interpretation of the Schrödinger equation in quantum mechanics from the control-theoretic point of view. Automatica. 2019 Jan;**1** (99):181-187

[12] Millet A, Morien PL. On a stochastic wave equation in two space dimensions: Regularity of the solution and its density. Stochastic Processes and Their Applications. 2000;**86**(1):141-162

[13] Ma J, Zhang G, Hayat T, Ren G. Model electrical activity of neuron under

electric field. Nonlinear Dynamics. 2019; **95**(2):1585-1598

[14] Li BK, Wen-Hsiung. Stochastic models in population genetics. Benchmark papers in genetics Vol. 7. Dowden, Hutchinson & Ross, Stroudsburg, Pennsylvania 1977. 484 S. Biometrical Journal [Internet]. 1979; **21**(3):297-297. Available from: http://doi.wiley.com/10.1002/bimj.4710210311

[15] Massabó M, Cianci R, Paladino O. An analytical solution of the advection dispersion equation in a bounded domain and its application to laboratory experiments. Journal of Applied Mathematics. 2011;**2011**:14. Article ID 493014. DOI: 10.1155/2011/493014

[16] Farmer WH, Vogel RM. On the deterministic and stochastic use of hydrologic models. Water Resource Research [Internet]. 2016;**52**(7): 5619-5633. Available from: https://onlinelibrary.wiley.com/doi/full/10.1002/2016WR019129

[17] Pang L, Hunt B. Solutions and verification of a scale-dependent dispersion model. Journal of Contaminant Hydrology. 2001;**53**(1–2):21-39

[18] Soheili AR, Soleymani F. Iterative methods for nonlinear systems associated with finite difference approach in stochastic differential equations. 2016;**71**:89-102

[19] Dereich S, Li S. Multilevel Monte Carlo implementation for SDEs driven by truncated stable processes. In: Springer Proceedings in Mathematics and Statistics. New York LLC: Springer; 2016. pp. 3-27

[20] Gantner RN, Schwab C. Computational higher order Quasi-Monte Carlo integration. In: Cools R, Nuyens D, editors. Monte Carlo and Quasi-Monte Carlo Methods. Springer Proceedings in Mathematics & Statistics. Vol. 163. Cham: Springer; 2016. DOI: 10.1007/978-3-319-33507-0_12

[21] Xiu D, Shen J. Efficient Stochastic Galerkin Methods for Random Diffusion Equations. Journal of Computational Physics. 2009;**228**(2):266-281. DOI: 10.1016/j.jcp.2008.09.008

[22] Matthies HG, Keese A. Galerkin methods for linear and nonlinear elliptic stochastic partial differential equations. Computer Methods in Applied Mechanics and Engineering. 2005;**194** (12–16):1295-1331

[23] Frauenfelder P, Schwab C, Todor RA. Finite elements for elliptic problems with stochastic coefficients. Computer Methods in Applied Mechanics and Engineering. 2005;**194** (2-5 SPEC. ISS):205-228

[24] Keese A. A Review of Recent Developments in the Numerical Solution of Stochastic Partial Differential Equations (Stochastic Finite Elements) Institut für Wissenschaftliches Rechnen. 2003;**06**. DOI: 10.24355/dbbs.084-200511080100-583200511080100-583

[25] Kaintura A, Dhaene T, Spina D. Review of polynomial chaos-based methods for uncertainty quantification in modern integrated circuits. Electronics (Basel) [Internet]. 2018;**7**(3): 30. Available from: http://www.mdpi.com/2079-9292/7/3/30

[26] Wan X, Karniadakis GE. Multi-element generalized polynomial chaos for arbitrary probability measures. SIAM Journal on Scientific Computing [Internet]. 2006;**28**(3):901-928. Available from: https://epubs.siam.org/doi/10.1137/050627630

[27] Kim KKK, Shen DE, Nagy ZK, Braatz RD. Wiener's polynomial chaos

for the analysis and control of nonlinear dynamical systems with probabilistic uncertainties. Control Systems. IEEE. 2013;**33**(5):58-67

[28] Kloeden PE, Platen E. Numerical Solution of Stochastic Differential Equations. Stochastic Modelling and Applied Probability. Springer Science & Business Media. 2011;**23**:636

[29] Zhang Z, Karniadakis GE. Numerical Methods for Stochastic Partial Differential Equations with White Noise. Vol. 1962017. Available from: http://link.springer.com/10.1007/978-3-319-57511-7

[30] Kroese DP, Taimre T, Botev ZI. Handbook of Monte Carlo Methods [Internet]2011. pp. 1-752. Available from: https://onlinelibrary.wiley.com/doi/book/10.1002/9781118014967

[31] Najm HN. Uncertainty Quantification and Polynomial Chaos Techniques in Computational Fluid Dynamics [Internet]. 2008. Available from: http://dx.doi.org/101146/annurev.fluid010908165248

[32] Xiu D, Em KG. The wiener-Askey polynomial chaos for stochastic differential equations. SIAM Journal on Scientific Computing [Internet]. 2003; **24**(2):619-644. Available from: http://epubs.siam.org/doi/10.1137/S1064827501387826

[33] Cameron RH, Martin WT. The orthogonal development of non-linear Functionals in series of Fourier-Hermite Functionals. The Annals of Mathematics. 1947;**48**(2):385

[34] Marelli S, Sudret B. UQLab: A Framework for Uncertainty Quantification in Matlab. 2014

[35] El-Amrani M, Seaid M, Zaidi NL. A new stochastic approach for advection-diffusion problems with uncertain parameters. Frontiers in Science and Engineering. 2012;**2**(1)

[36] HIDA T, IKEDA N. Analysis on Hilbert space with reproducing kernel arising from multiple wiener integral. Selected Papers of Takeyuki Hida. 2001: 142-168

[37] Holden H. Stochastic Partial Differential Equations: A Modeling, White Noise Functional Approach. 1996. p. 230

[38] Hu Y, Jan Y. Wick Calculus for Nonlinear Gaussian Functionals. 2009: arXiv:0901.4911v1. DOI: 10.48550/arXiv.0901.4911

[39] Holden H, Lindstrøm T, Øksendal B, Ubøe J, Zhang T. Stochastic Boundary Value Problems. A White Noise Functional Approach. 1991. Available from: https://www.duo.uio.no/handle/10852/43486

Chapter 4

Some Remarks on Vector Markov Chains and Their Applications to the Description of Many-Particle Systems

Bruno Carbonaro and Federica Vitale

Abstract

Although in the literature on the topic, this point is never expressed, but regularly implied, an important and widely studied scheme of models to describe and predict, at least in stochastic terms, the behavior of many-particle systems is based on Markov Chains. In fact, Markov Chains are special stochastic processes with important properties that have revealed to be shared with many natural processes. This fact, though almost obvious, requires to be carefully discussed, at least to point out that the very notion of a Markov Chain must be completed and somehow generalized to understand the way in which Markov Chains intervene in the equations describing the stochastic behavior of any system. This is not a purely critical and "foundational" task but aims to develop a new very plausible and likely way to introduce possible effects of external world on a non-isolated many-particle system. This paper aims to offer at least a basic discussion of this perspective.

Keywords: stochastic processes, Markov Chains, probability, kinetic theory, mathematical models, complex systems

1. Introduction

The aim of the present chapter is twofold.

On one hand, we want to consider explicitly some properties of a kind of Markov Chains [1–3], which is of course taken implicitly into account in the whole literature about these important stochastic processes but—as far as we are aware—has been never examined explicitly, that is, the Markov Chains whose states are n-tuples of integers or of real numbers, which can be called *vector* Markov Chains. In particular, we will also pay attention in particular to nonstationary Markov Chains, which we recall in some details in Section 1.2.

On the other hand, we want to examine the equations of kinetic-theoretic models [4] that since almost forty years seem to be diffusing ever more in the research about the evolution of many-particle systems and appear to be a particularly effective, expressive, and versatile tool to describe in stochastic terms such evolution [5–10].

IntechOpen

Rather strange to say, these equations are firmly based on the use of Markov Chains, since in them a basic rôle is played by transition matrices, that just characterize these important stochastic processes, but in the literature about the kinetic-theoretic models Markov Chains are never even mentioned (at least to our knowledge). The researchers in this field probably consider this reference superfluous. But the question spontaneously arises whether if we exploit the rôle of Markov Chains then our understanding of the terms of equations could be improved, suggesting to us some modifications that could help us to produce a good and effective description of the behavior of systems that have till now escaped the classical equations used up to the last few years (in this connection, see [11, 12], where an accurate stochastic description of the evolution of a system is obtained renouncing the Principle of Inertia, and [13–15] where new perspectives about the description of non-isolated systems have been obtained by introducing «forcing» terms in the equations).

Accordingly, after giving some definitions recalling some types of Markov Chains and introducing explicitly the *vector* Markov Chains, of which we shall discuss some basic properties that could be useful to understand the rôle of transition matrices in the equations of the kinetic-theoretic models, we shall first recall the structure of these latter, giving a critical description of the terms involved therein, and finally, we will suggest the ways in which, in virtue of our knowledge of the complete meaning of the transition probabilities, and in view of catching new natural processes for non-isolated systems, the transition probabilities (together with the so-called encounter rates, see Sections 1.6 and 1.7) can be modified.

2. Some background about Markov Chains

As is well-known, a random process [2] is a sequence $\{X_\eta\}_{\eta\in\mathbb{H}}$ of random variables with a common range S of possible values (the *states*) called the *state space* of the process. Both $\mathbb{H}$ and S can be either discrete or continuous sets. If $\mathbb{H}$ is discrete, that is $\mathbb{H}=\mathbb{N}=\{0,1,2, \ \dots \ ,n, \ \dots\}$, then the process is said to be *time-discrete*, while, if $\mathbb{H}$ is continuous, that is, $\mathbb{H}=[0,+\infty)$, then the process is said to be *time-continuous*. Analogously if S is discrete, that is, $S=\{z_h\}_{h\in\mathbb{I}}\subseteq\mathbb{R}$ with $\mathbb{I}\subseteq\mathbb{Z}$, then the process is said to be *discrete*, while, if S is continuous (it is any real interval), then the process is said to be *continuous*. From now to the end of this section, we shall first consider discrete and time-discrete processes, and at a second step time-continuous discrete processes. In both cases, for the sake of simplicity, the state space S will be assumed to be finite. In this choice, there is no loss of generality, and we only avoid to write matrices with infinitely many rows and columns.

Consider then a discrete and time-discrete random process $\{X_h\}_{h\in\mathbb{N}}$ with $S=\{z_1,z_2, \dots ,z_n\}$. . As usual, we shall denote by a vector $\mathbf{p}_h$ (called the *state vector of the process at time h*) the (absolute) probability distribution on S according to the random variable X_h. In other words, $\mathbf{p}_h\equiv\left(p_{h,1},p_{h,2}, \dots ,p_{h,n}\right)$ with $p_{h,k}=\mathbf{P}(X_h=z_k)$.[1]

Now, a discrete time-discrete Markov Chain $\{X_h\}_{h\in\mathbb{N}}$, with $R(X_h)=S\subset\mathbb{Z}$, is a random process such that the Markov condition

[1] Here and in the sequel, any probability measure, when explicitly applied to events, will be denoted by a bold-face capital letter.

$$\begin{aligned}\mathbf{P}(X_h = z_h \,|X_1 = z_1, X_2 = z_2, \ \dots \ , X_{h-1} = z_{h-1}) = \\ = \mathbf{P}(X_h = z_h \,|X_{h-1} = z_{h-1}), \qquad \forall h \in \mathbb{N}\backslash\{0\}\end{aligned} \tag{1}$$

holds (where we have used the usual symbol $\mathbf{P}(A\,|B)$ to denote the conditional probability of an event A under the assumption that an event B has taken place). This means that, for any $h \in \mathbb{N}$, the values of the h-th random variable of the process depend only on the values of the $(h-1)$-th, not on the values of any previous variable. Now, setting $P_{ij}(h) = \mathbf{P}(X_h = z_j\,|X_{h-1} = z_i)$ (with $(i,j) \in \{1, 2, \ \dots \ , n\}^2$), $P_{ij}(h)$ is the *transition probability* from state z_i to state z_j at the h-th step, and the matrix

$$\mathbf{P}_h \equiv \left(P_{ij}(h)\right)_{1\le i,j\le n} \equiv \begin{pmatrix} P_{11}(h) & P_{12}(h) & \dots & P_{1n}(h) \\ P_{21}(h) & P_{22}(h) & \dots & P_{2n}(h) \\ \vdots & \vdots & \ddots & \vdots \\ P_{n1}(h) & P_{n2}(h) & \dots & P_{nn}(h) \end{pmatrix}, \qquad \forall h \in \mathbb{N}\backslash\{0\}. \tag{2}$$

is called the *transition matrix* at h-th step. It is well-known (and also obvious) that

$$\sum_{j=1}^{n} P_{ij}(h) = 1, \qquad \forall h \in \mathbb{N}\backslash\{0\} \tag{3}$$

and, in virtue of the law of alternatives,

$$\mathbf{p}_h = \mathbf{p}_{h-1}\mathbf{P}_h, \qquad \forall h \in \mathbb{N}\backslash\{0\}, \tag{4}$$

where $\mathbf{p}_h$ is a row vector for any h, so that the product at right-hand side is a row-by-column product. For the sake of completeness, we also recall that, for any couple (r, s) of nonnegative integers,

$$\mathbf{p}_{r+s} = \mathbf{p}_r\mathbf{P}_{r+1}\mathbf{P}_{r+2} \dots \mathbf{P}_{r+s}, \tag{5}$$

and in particular, when the Markov Chain is stationary, that is, a transition matrix $\mathbf{P}$ exists such that $\mathbf{P}_h = \mathbf{P}$ for any $h \in \mathbb{N}\backslash\{0\}$,

$$\mathbf{p}_{r+s} = \mathbf{p}_r\mathbf{P}^s, \tag{6}$$

where the power at right-hand side must be interpreted in the sense of the row-by-column product of matrices (for further details, see e.g. [2, 3]).

3. Joint and marginal transition probabilities

Now, bearing in mind these premises, we want to observe that in the literature about Markov Chains, the structure of the state space S is usually completely disregarded. With only one exception at our knowledge [16], it is simply assumed to be an ordered set, mainly to be allowed to write simple transition matrices. But we want to ask ourselves which new features should be taken into account if we considered some kinds of structure, and how should develop the study. In particular, we want to examine the case in which $S = \Omega^m$ (where $\Omega = \{x_1, x_2, \ \dots \ , x_l\}$ so that

$|\Omega^m| = l^m$ and $\mathbf{z}_i \equiv (x_{i_1}, x_{i_2}, \ldots, x_{i_m})$ for any $i \in \{1, 2, \ldots, n\}$ and $(i_1, i_2, \ldots, i_m) \in \{1, 2, \ldots, l\}^m)$ and the process on consideration should be described as a sequence $\{(X_{h,1}, X_{h,2}, \ldots, X_{h,m})\}_{h \in \mathbb{N}}$, which will be written in vector notation as $\{\mathbf{X}_h\}_{h \in \mathbb{N}}$. In such a case, we should consider the transition probabilities

$$\begin{aligned} &\mathbf{P}(\mathbf{X}_h = \mathbf{z}_h \mid \mathbf{X}_{h-1} = \mathbf{z}_{h-1}) = \\ &\quad = \mathbf{P}(X_{h,1} = x_{h,1}, \ldots, X_{h,m} = x_{h,m} | X_{h-1,1} = x_{h-1,1}, \ldots, X_{h-1,m} = x_{h-1,m}) \end{aligned} \tag{7}$$

that have $2m$ indexes, that is, we can set

$$\begin{aligned} &\mathbf{P}(X_{h,1} = x_{j_1}, \ldots, X_{h,m} = x_{j_m} | X_{h-1,1} = x_{i_1}, \ldots, X_{h-1,m} = x_{i_m}) \equiv \\ &\quad \equiv P_{i_1, i_2, \ldots, i_m, j_1, j_2, \ldots, j_m}(h). \end{aligned} \tag{8}$$

So, for any $h \in \mathbb{N}$, the transition matrix $\mathbf{P}_h$ at h-th time is a $2m$-dimensional matrix. It will be called the *joint transition matrix*, and each of its elements $P_{i_1, i_2, \ldots, i_m, j_1, j_2, \ldots, j_m}(h)$ expresses the probability that the chain passes from the state $(x_{i_1}, \ldots, x_{i_m})$ to the state $(x_{j_1}, \ldots, x_{j_m})$ at h-th time. These will be called the *joint transition probabilities*.

But, of course, together with these joint probabilities, we are now allowed to consider also the many different *marginal probabilities*. More precisely, for any $h \in N$, we can choose $r < m$ of the random variables $X_{h,1}, X_{h,2}, \ldots, X_{h,m}$, say $X_{h,k_1}, X_{h,k_2}, \ldots, X_{h,k_r}$ and look for the probabilities $\mathbf{P}\left(X_{h,k_1} = x_{j_{k_1}}, \ldots, X_{h,k_r} = x_{j_{k_r}} | X_{h-1,1} = x_{i_1}, \ldots, X_{h-1,m} = x_{i_m}\right)$. For any choice, we have a $(r+m)$-dimensional matrix with l^{r+m} entries. Of course, the matrices are $\binom{n}{r}$. We shall denote by $P_{i_1, i_2, \ldots, i_m, j_{k_1}, j_{k_2}, \ldots, j_{k_r}}(h)$ the entries of each of such matrices. It should also be noted that, setting $s = m - r$, and agreeing to denote by $X_{h,\ell_1}, X_{h,\ell_2}, \ldots, X_{h,\ell_s}$ the remaining s variables, we have

$$\boldsymbol{P}_{i_1, i_2, \ldots, i_m, j_{k_1}, j_{k_2}, \ldots, j_{k_r}}(h) = \sum_{j_{\ell_1}, \ldots, j_{\ell_s}}^{1 \ldots n} \boldsymbol{P}_{i_1, i_2, \ldots, i_m, j_{k_1}, \ldots, j_{k_r}, j_{\ell_1}, \ldots, j_{\ell_s}}(h) \tag{9}$$

Now, for the sake of simplicity, assume that $k_1 = 1, k_2 = 2, \ldots, k_r = r$ and $\ell_1 = r+1, \ell_2 = r+2, \ldots, \ell_s = m$. Then, we shall also agree to set $S = \{\mathbf{z}_{ij}\}_{\substack{1 \le i \le p \\ 1 \le j \le q}} \equiv \Omega^r \times \Omega^s \equiv \left\{\left(\boldsymbol{x}_i, \boldsymbol{y}_j\right)\right\}_{\substack{1 \le i \le p \\ 1 \le j \le q}}$ $(p = l^r, q = l^s)$, where $r + s = m$, $x_i \equiv \left(x_i^1, x_i^2, \ldots, x_i^r\right)$ and $\boldsymbol{y}_j \equiv \left(x_j^{r+1}, x_j^{r+2}, \ldots, x_j^m\right)$[2]. Then, we can write relation (9) in the form

$$P_{\mathbf{z}_{hk}, \boldsymbol{x}_i}(h) = \sum_{j=1}^{q} P_{\mathbf{z}_{hk}, \mathbf{z}_{ij}}(h) = \sum_{j=1}^{q} P_{\mathbf{z}_{hk}, \left(\boldsymbol{x}_i, \boldsymbol{y}_j\right)}(h), \tag{10}$$

[2] From now on, the symbols $\mathbf{z}_{ij}$ and $\left(\boldsymbol{x}_i, \boldsymbol{y}_j\right)$ will be treated as interchangeable. Moreover, the indexes of components of any vector will be written as apices for the sake of readability.

and introduce the joint state vector $\mathbf{p}_h \equiv \left(p_h^{ij}\right)$, with $p_h^{ij} \equiv p_h(\mathbf{z}_{ij}) \equiv p_h\left(\boldsymbol{x}_i, \boldsymbol{y}_j\right)$, and the marginal state vector $\mathbf{p}_{x,h} \equiv \left(p_{x,h}(\boldsymbol{x}_i)\right)_{1 \le i \le p}$, where

$$p_{x,h}(\boldsymbol{x}_i) = \sum_{j=1}^{q} p_h(\mathbf{z}_{ij}) = \sum_{j=1}^{q} p_h\left(\boldsymbol{x}_i, \boldsymbol{y}_j\right) \tag{11}$$

and, in virtue of the law of alternatives

$$p_h(\mathbf{z}_{ij}) = \sum_{k=1}^{p} \sum_{l=1}^{q} p_{h-1}(\mathbf{z}_{kl}) P_{\mathbf{z}_{kl}, \mathbf{z}_{ij}}(h).$$

Hence, replacing this last relation in relation (11), and taking into account relation (10), we get

$$\begin{aligned} p_{x,h}(\boldsymbol{x}_i) &= \sum_{j=1}^{q} \sum_{k=1}^{p} \sum_{l=1}^{q} p_{h-1}(\boldsymbol{x}_k, \boldsymbol{y}_l) P_{\mathbf{z}_{kl}, (\boldsymbol{x}_i, \boldsymbol{y}_j)}(h) = \\ &= \sum_{k=1}^{p} \sum_{l=1}^{q} p_{h-1}(\boldsymbol{x}_k, \boldsymbol{y}_l) \left(\sum_{j=1}^{q} P_{\mathbf{z}_{kl}, (\boldsymbol{x}_i, \boldsymbol{y}_j)}(h) \right) = \\ &= \sum_{k=1}^{p} \sum_{l=1}^{q} p_{h-1}(\boldsymbol{x}_k, \boldsymbol{y}_l) P_{\mathbf{z}_{kl}, \boldsymbol{x}_i}(h). \end{aligned}$$

Now, for reasons that will be clear in the sequel, we write this last equation in the form

$$p_{x,h}(\boldsymbol{x}_i) - p_{x,h-1}(\boldsymbol{x}_i) = \sum_{k=1}^{p} \sum_{l=1}^{q} p_{h-1}(\mathbf{z}_{kl}) P_{\mathbf{z}_{kl}, \boldsymbol{x}_i}(h) - p_{x,h-1}(\boldsymbol{x}_i). \tag{12}$$

In addition, for any $l \in \{1, 2, \dots, q\}$,

$$\sum_{i'=1}^{p} \sum_{l'=1}^{q} P_{(\boldsymbol{x}_i, \boldsymbol{y}_l), (\boldsymbol{x}_{i'}, \boldsymbol{y}_{l'})}(h) = 1,$$

so that we can rewrite equation (12) in the final form

$$p_{x,h}(\boldsymbol{x}_i) - p_{x,h-1}(\boldsymbol{x}_i) = \sum_{l=1}^{q} \left\{ \sum_{k \ne i}^{1\dots p} p_{h-1}(\mathbf{z}_{kl}) P_{\mathbf{z}_{kl}, \boldsymbol{x}_i}(h) + - p_{h-1}(\mathbf{z}_{il}) \sum_{i' \ne i}^{1\dots p} P_{(\boldsymbol{x}_i, \boldsymbol{y}_l), \boldsymbol{x}_{i'}}(h) \right\}. \tag{13}$$

Finally, if for any $h \in \mathbb{N}$, the random variables $X_{h,i}$ and $X_{h,j}$ are independent for any couple (i, j), and the chain is stationary, then we find

$$p_{x,h}(\boldsymbol{x}_i) - p_{x,h-1}(\boldsymbol{x}_i) = \sum_{l=1}^{q} \left\{ \sum_{k \ne i}^{1\dots p} p_{x,h-1}(\boldsymbol{x}_k) p_{y,h-1}(\boldsymbol{y}_l) P_{\mathbf{z}_{kl}, \boldsymbol{x}_i} + - p_{x,h-1}(\boldsymbol{x}_i) \sum_{k \ne i}^{1\dots p} p_{y,h-1}(\boldsymbol{y}_l) P_{\mathbf{z}_{il}, \boldsymbol{x}_k} \right\}. \tag{14}$$

The first term at right-hand side will be called the *gain term* of the subset of states $S_i = \{\mathbf{z}_{il}\}_{1 \le l \le q}$ since it accounts for all the possible transitions from other states to a state of S_i, while the last term is called the *loss term* of S_i since it accounts for all the possible transitions from a state of S_i to any state *outside* S_i.

4. Time-continuous vector Markov Chains

We now want to turn our attention to the case of time-continuous Markov Chains. But before treating that case, we want first to consider the random process $\{(\mathbf{X}_h, \chi_h)\}_{h\in\mathbb{N}}$, where, for any $h \in \mathbb{N}$, χ_h is the classical Bernoulli variable, with range $R(\chi_h) = \{0, 1\}$. The variables χ_h are independent and identically distributed, with $\mathbf{P}(\chi_h = 0) = v$ and $\mathbf{P}(\chi_h = 1) = u$ for any $h \in \mathbb{N}$. For any $h \in \mathbb{N}$, $\mathbf{X}_h$, the state vector $\mathbf{p}_h$ and the transition matrix $\mathbf{P}_h$ are assumed to depend only on χ_h. So, using again the notation introduced above, we see at once that
$p_h(\mathbf{z}_{ij}) = \mathbf{P}(\mathbf{X}_h = \mathbf{z}_{ij}, \chi_h = 1) + \mathbf{P}(\mathbf{X}_h = \mathbf{z}_{ij}, \chi_h = 0)$, that is,

$$p_h(\mathbf{z}_{ij}) = \mathbf{P}(\mathbf{X}_h = \mathbf{z}_{ij} \,|\, \chi_h = 1)u + \mathbf{P}(\mathbf{X}_h = \mathbf{z}_{ij} \,|\, \chi_h = 0)v. \tag{15}$$

Analogously, for any $h \in \mathbb{N}$, the transition matrix at h-th step (or time) should be written according to the condition

$$P_{\mathbf{z}_{ij},\mathbf{z}_{kl}}(h) = P_{\mathbf{z}_{ij},\mathbf{z}_{kl}\,|\,\chi_h=1}(h)u + P_{\mathbf{z}_{ij},\mathbf{z}_{kl}\,|\,\chi_h=0}(h)v,$$

where $P_{\mathbf{z}_{ij},\mathbf{z}_{kl}\,|\,\chi_h=1}(h)$ is the transition probability from the state $\mathbf{z}_{ij}$ to the state $\mathbf{z}_{kl}$ at h-th step under the assumption $\chi_h = 1$, and $P_{\mathbf{z}_{ij},\mathbf{z}_{kl}\,|\,\chi_h=0}(h)$ is the transition probability from the state $\mathbf{z}_{ij}$ to the state $\mathbf{z}_{kl}$ at h-th step under the assumption $\chi_h = 0$. In particular, we shall set

$$P_{\mathbf{z}_{ij},\mathbf{z}_{kl}\,|\,\chi_h=0}(h) = \delta^i_k \delta^j_l, \qquad \forall h \in \mathbb{N}: \tag{16}$$

this means that the chain remains in the state occupied at time h and also at time $h + 1$ when $\chi_h = 0$. Also, notice that the Markov Chain is stationary if and only if the matrix $P_{\mathbf{z}_{ij},\mathbf{z}_{kl}\,|\,\chi_h=1}$ is constant with respect to h.

By replacing the well-known relations

$$\mathbf{P}(\mathbf{X}_{h+1} = \mathbf{z}_{ij} \,|\, \chi_h = 1) = \sum_{k=1}^{p} \sum_{l=1}^{q} p_h(\mathbf{z}_{kl}) P_{\mathbf{z}_{kl},\mathbf{z}_{ij}\,|\,\chi_h=1}(h),$$
$$\mathbf{P}(\mathbf{X}_{h+1} = \mathbf{z}_{ij} \,|\, \chi_h = 0) = \sum_{k=1}^{p} \sum_{l=1}^{q} p_h(\mathbf{z}_{kl}) P_{\mathbf{z}_{kl},\mathbf{z}_{ij}\,|\,\chi_h=0}(h),$$

in equation (15), and taking into account assumption (16) and the obvious relation $v = 1 - u$, we get

$$p_{h+1}(\mathbf{z}_{ij}) - p_h(\mathbf{z}_{ij}) = \sum_{k=1}^{p} \sum_{l=1}^{q} p_h(\mathbf{z}_{kl}) P_{\mathbf{z}_{kl},\mathbf{z}_{ij}\,|\,\chi_h=1}(h)u - p_h(\mathbf{z}_{ij})u. \tag{17}$$

These are of course pq equations, one for any couple (x_i, y_j). By taking the sum over j of both sides, we obtain the following system of p equations:

$$p_{\boldsymbol{x},h+1}(\boldsymbol{x}_i) - p_{\boldsymbol{x},h}(\boldsymbol{x}_i) = \\ = \sum_{k=1}^{p}\sum_{l=1}^{q} p_h(\mathbf{z}_{kl}) P_{\mathbf{z}_{kl},\boldsymbol{x}_i|\chi_h=1}(h)u - p_{\boldsymbol{x},h}(\boldsymbol{x}_i)u, \qquad (i = 1, 2, \dots, p). \tag{18}$$

Finally, it is readily seen that, under the independence assumption on the random variables $X_{h,i}$ and $X_{h,j}$ for any h, we obtain the system

$$p_{\boldsymbol{x},h+1}(\boldsymbol{x}_i) - p_{\boldsymbol{x},h}(\boldsymbol{x}_i) = \\ = u\sum_{l=1}^{q}\left\{\sum_{k\neq i}^{1\dots p} p_{\boldsymbol{x},h}(\boldsymbol{x}_k)p_{\boldsymbol{y},h}(\boldsymbol{y}_l)P_{\mathbf{z}_{kl},\boldsymbol{x}_i|\chi=1} + \\ -p_{\boldsymbol{x},h}(\boldsymbol{x}_i)\sum_{k\neq i}^{1\dots p} p_{\boldsymbol{y},h}(\boldsymbol{y}_l)P_{\mathbf{z}_{il},\boldsymbol{x}_k|\chi=1}\right\}, \qquad (i = 1, 2, \dots, p). \tag{19}$$

if the transition matrix $\mathbf{P}_{\chi=1} \equiv \left(P_{\mathbf{z}_{ij},\mathbf{z}_{kl}|\chi=1}\right)$ is stationary.

We are now ready to consider the case of a time-continuous Markov Chain analogous to the one; we have just described, namely a random process of the type $\{(\mathbf{X}_t, \chi_t)\}_{t\in[0,T)}$, where $T > 0$ may be either finite or infinite. In this case, the state vectors obviously depend on time, but the transition probabilities may depend on time or not, as well as the probabilities associated to the Bernoulli variables χ_t. But, similarly to what has been done above, these variables will be assumed to be identically distributed. In this connection, we have to recall to the reader that it is impossible to assign a positive probability u such that $\mathbf{P}(\chi_t = 1) = u$ for any t because in such case the probability that at least one of the variables χ_t could take the value 1 for t in any interval contained in $[0, T)$ would be infinite. What we can do is to assign a *constant* probability density τ such that $\tau\Delta t$ is the probability to find in any interval of length Δt some points such that $\chi_t = 1$. Roughly speaking and referring to the interpretation of probability as relative frequency, we can state that, for any $s \in [0, T)$, $\tau\Delta t$ is the length of the set $\{t \in [s, s + \Delta t] \mid \chi_t = 1\}$ so that τ is the ratio of that length to the length Δt. Accordingly, τ is the measure of the set of points for which $\chi_t = 1$ in any interval of unit length.

This stated, we shall write system (19) as follows:

$$p_{\boldsymbol{x},t+\Delta t}(\boldsymbol{x}_i) - p_{\boldsymbol{x},t}(\boldsymbol{x}_i) = \\ = \tau\Delta t\sum_{l=1}^{q}\left\{\sum_{k\neq i}^{1\dots p} p_{\boldsymbol{x},t}(\boldsymbol{x}_k)p_{\boldsymbol{y},t}(\boldsymbol{y}_l)P_{\mathbf{z}_{kl},\boldsymbol{x}_i|\chi=1} + \\ -p_{\boldsymbol{x},t}(\boldsymbol{x}_i)\sum_{k\neq i}^{1\dots p} p_{\boldsymbol{y},t}(\boldsymbol{y}_l)P_{\mathbf{z}_{il},\boldsymbol{x}_k|\chi=1}\right\}. \qquad (i = 1, 2, \dots, p), \tag{20}$$

where $\tau\Delta t$ replaces the probability u in this continuous framework.

Next, dividing both sides by Δt and letting $\Delta t \to 0$, we get

$$\frac{p_{\boldsymbol{x},t}}{t}(\boldsymbol{x}_i) = \tau\sum_{l=1}^{q}\left\{\sum_{k\neq i}^{1\dots p} p_{\boldsymbol{x},t}(\boldsymbol{x}_k)p_{\boldsymbol{y},t}(\boldsymbol{y}_l)P_{\mathbf{z}_{kl},\boldsymbol{x}_i|\chi=1} + \\ -p_{\boldsymbol{x},t}(\boldsymbol{x}_i)\sum_{k\neq i}^{1\dots p} p_{\boldsymbol{y},t}(\boldsymbol{y}_l)P_{\mathbf{z}_{il},\boldsymbol{x}_k|\chi=1}\right\}. \qquad (i = 1, 2, \dots, p). \tag{21}$$

We will show that the above «mixed» process we have just discussed above, obtained by coupling a Markov Chain with a uniform process, can be further generalized, by taking into account *vector* uniform processes of an appropriate dimension. But we will do this in a different context, namely in the context of the models of collective phenomena.

5. Some background about the stochastic description of many-particle systems

For reasons that will be clear in a very short, while we shall now focus our attention on the double indexing of the elements on S, and assume $r = s$ is so that $m = 2s$, $S = \Omega^{2s}$, $p = q = l^s$ and $n = l^{2s}$ and consider the random process $\{(\mathbf{X}_t, \chi_t)\}_{t\in[0,T)}$, where $0 < T \leq +\infty$ and $\chi_t \equiv \left(\chi_t^{ij}\right)_{1\leq i,j\leq l^s}$ for any $t \in [0, T)$. According to our notation conventions, χ_t is in fact for any t a vector with n components that are double-indexed just to reproduce the (conventional) double indexing of the elements of S. We can treat it as a square matrix with χ_t rows and l^s columns. For each entry χ_t^{ij} of this matrix, we shall set

$$\chi_t^{ij} \equiv \{\chi_t^{kl} \in \chi_t \mid \text{either } \ k \neq i \ \text{ or } \ l \neq j\}.$$

For any t and for any couple (i, j), χ_t^{ij} is a Bernoulli random variable, and we assume that

1. For any t and t' in $[0, T)$, χ_t^{ij} and $\chi_{t'}^{ij}$ are independent and identically distributed. Accordingly, $\mathbf{P}\left(\chi_t^{ij} = 1\right) = u_{ij}$ and $\mathbf{P}\left(\chi_t^{ij} = 0\right) = 1 - u_{ij}$ for any $t \in [0, T)$.

2. For any $t \in [0, T)$ and for any $(i,j) \in \{1, 2, \ldots, l^s\}^2$, the probability of state $\mathbf{z}_{ij}$ at time t depends only on χ_t^{ij}.

3. For any $t \in [0, T)$ and for any $(i,j) \in \{1, 2, \ldots, l^s\}^2$, the transition probabilities from the state $\mathbf{z}_{ij}$ to any other state depend only on χ_t^{ij}; accordingly, we are allowed to consider only the conditional transition probabilities $P_{\mathbf{z}_{ij},\mathbf{z}_{kl}|\chi_t^{ij}=0}(t)$ and $P_{\mathbf{z}_{ij},\mathbf{z}_{kl}|\chi_t^{ij}=1}(t)$.

4. As for the process considered in the previous section,

$$P_{\mathbf{z}_{ij},\mathbf{z}_{kl}|\chi_t^{ij}=0}(t) = \delta_k^i \delta_l^j, \qquad \forall t \in [0, T),$$

so that we are allowed again to state that the chain is stationary if and only if $P_{\mathbf{z}_{ij},\mathbf{z}_{kl}|\chi_t^{ij}=1}(t)$ is in fact independent of t.

This stated that we are allowed to reproduce the reasonings of the previous sections, with obvious changes leading us to replace the system (21) with the system

$$\begin{aligned}\frac{p_{\boldsymbol{x},t}}{t}(\boldsymbol{x}_i) = \sum_{l=1}^{q}\Bigg\{&\sum_{k\neq i}^{1\dots p}\tau_{kl}p_{\boldsymbol{x},t}(\boldsymbol{x}_k)p_{\boldsymbol{y},t}(\boldsymbol{y}_l)P_{\mathbf{z}_{kl},\boldsymbol{x}_i\,|\,\chi_t^{kl}=1}+\\ &-p_{\boldsymbol{x},t}(\boldsymbol{x}_i)\sum_{k\neq i}^{1\dots p}\tau_{il}p_{\boldsymbol{y},t}(\boldsymbol{y}_l)P_{\mathbf{z}_{il},\boldsymbol{x}_k\,|\,\chi_t^{il}=1}\Bigg\}, \qquad (i = 1, 2, \dots, l^s).\end{aligned} \tag{22}$$

Now, once we have reconstructed the way in which this last system of equations is originated by coupling a stationary vector Markov Chain with an additional vector random process actually modifying the final form of the transition matrix maintaining; however, its stationarity (that is its independence on time), we shall devote the rest of this section to recall a classical interpretation (see [8]) of system (22), which has become in the last forty years ever more important as it can be applied to a wide range of different phenomena, from mechanics of gases (which is its true historical origin, since Boltzmann proposed his celebrated kinetic theory [4]) to the behavior of biological systems, with particular concern to the interaction between healthy and tumor tissues (see [6], also for more complete bibliographic references), from social sciences, with particular concern to the diffusion of competing opinions [14], to economics [9], and to the behavior of swarms and crowds [5, 7]. This interpretation has in fact given rise to a general model for the description of the behavior of a large number of many-particle systems, we would dare say of almost all possible types and in almost all possible contexts.

The key idea upon which the kinetic-theoretic description of the behavior of many-particle systems is based and is the following. A set $\mathcal{S}$ of a very large number N of objects is given. The objects of $\mathcal{S}$, usually called «particles» or «individuals» according to the context, can interact pairwise[3] with each other and are identified (further than by some common «physical» properties they can possess to different degrees, which do not appear explicitly in the mathematical description of the behavior of $\mathcal{S}$, but influence the values of parameters) by their «states». According to the context, the set of possible states of all particles can be finite, countably infinite, or continuous. We cannot (and do not) claim to know exactly the state of each particle of the system at any time so that even the prescription of N precise initial conditions on the states of particles would be quite unrealistic. What we assume to be allowed to state is that, if the number of all possible different states is q, then all the particles of $\mathcal{S}$ can be thought of as divided into n different classes $\mathcal{S}_1, \mathcal{S}_2, \dots, \mathcal{S}_q$, each containing all and only the particles sharing the same state; if $n_i(t)$ (with $i \in \{1, 2, \dots n\}$ is the number of particles sharing the i-th state at time t, then $p_i(t) = n_i(t)/N$ is the probability to pick at random in $\mathcal{S}$ a particle in the i-th state, and we can construct a state vector $\mathbf{p} \equiv \left(p_1, p_2, \dots, p_q\right)$. In addition, each particle belonging to a class $\mathcal{S}_k$ can jump into another class $\mathcal{S}_i$ if and only if it «interacts» with some other particle: the state (that is, the class $\mathcal{S}_l^*$) of this latter does not matter, if not as influencing the assessment of the different probabilities of jump $P_{ki} = \mathbf{P}(\mathcal{S}_k, \mathcal{S}_i)$, that accordingly are denoted $P_{ki;l} = \mathbf{P}(\mathcal{S}_k, \mathcal{S}_i; \mathcal{S}_l^*)$. Moreover, if a particle has no interactions with other particles, then it will not change its state.

The above almost purely qualitative descriptions lead us at once to conclude that, in the above scheme, any many-particle system is *in fact* a random process $\{(\mathbf{X}_t, \chi_t)\}_{t\in[0,T)}$ of the above-described type, where $\{\mathbf{X}_t\}$ is a Markov Chain,

---

[3] But the theory is currently developing to cover the case of *multiple* interactions.

characterized by a state space, which in our previous description is $\Omega^s \equiv \{\boldsymbol{x}_1, \boldsymbol{x}_2, \dots, \boldsymbol{x}_q\}$ (where, we recall, $q = l^s$), a state (probability) vector $\mathbf{p}_x$ expressing in percentage the distribution of particles over all possible different states, an interaction rate τ_{ij} (the average number of interactions between a particle of $\boldsymbol{\mathcal{S}}_i$ and a particle of $\boldsymbol{\mathcal{S}}_j$ occurring in any unit time interval, that is, the measure of the set $I_1 \equiv \left\{t \in I \mid \chi_t^{ij} = 1\right\}$ in any inerval I of unit length), and each probability of jump $P_{ki;l} = \mathbf{P}\left(\boldsymbol{\mathcal{S}}_k, \boldsymbol{\mathcal{S}}_i; \boldsymbol{\mathcal{S}}_l^*\right)$, conditional to the occurrence of an interaction between a particle in the state $\boldsymbol{x}_k$ and a particle in the state $\boldsymbol{y}_l$ is given by $P_{(\boldsymbol{x}_k, \boldsymbol{y}_l), \boldsymbol{x}_i \mid \chi_t^{kl}=1} = P_{\mathbf{z}_{kl}, \boldsymbol{x}_i \mid \chi_t^{kl}=1}$. So, relations (22) turn out to be exactly the equations governing, in stochastic terms, the evolution of *isolated* many-particle systems.

6. Equations for many-particle systems with time-dependent transition matrices

At a first glance, the above reconstruction of the equations of kinetic theory in terms of a random process $\{(\mathbf{X}_t, \chi_t)\}_{t \in [0,T)}$ coupling a stationary Markov Chain and a Bernoullian process could seem a purely academic exercise, though interesting as showing the correspondence between two different languages, born in different contexts, with different aims and in different times[4]. But our exercise is not an end in itself but is very useful to extend our model for many-particle systems to cover the case of systems interacting with the external world. It is in fact not by chance that we have stressed that equations (22) hold for *isolated* many-particle systems. These equations are based on the assumption that the instantaneous variation of the distribution of particles over the state space is uniquely due to mutual interactions between particles and that these interactions modify the states of the involved particles always in the same way, independently of the time at which they occur[5]. But when the system perceives the influence of the external world, we are allowed to assume that both the interaction rate and the results of any interaction are more or less strongly modified from time to time by events that happen outside the system.

More precisely, our stochastic process $\{(\mathbf{X}_t, \chi_t)\}_{t \in [0,T)}$ must be replaced by another process $\{(\mathbf{X}_t, \chi_t, \mathbf{E}_t)\}_{t \in [0,T)}$, where, for any $t \in [0, T)$, $\mathbf{E}_t$ is another «strange» vector random variable[6] with values in $\mathbb{W} \equiv \mathbb{R}^d \times \mathbb{K}^h$ (where $\mathbb{K}$ is a finite subset of $\mathbb{Z}$), such that, for any $t \in [0, T)$, there exists an $\mathbf{e}(t) = (\mathbf{u}(t), \mathbf{v}(t)) \equiv (u_1(t), u_2(t), \dots, u_d(t), v_1(t), v_2(t), \dots, v_h(t)) \in \mathbb{W}$ $[\mathbf{u}(t) \equiv (u_1(t), u_2(t), \dots, u_d(t)), \mathbf{v}(t) \equiv (v_1(t), v_2(t), \dots, v_h(t))]$ for which the probability density function of $\mathbf{E}_t$ is expressed, for any $\boldsymbol{\varepsilon} \equiv (\boldsymbol{\xi}, \boldsymbol{\eta}) \equiv (\xi_1, \dots, \xi_d, \eta_1, \dots, \eta_h)$, with $\boldsymbol{\xi} \equiv (\xi_1, \dots, \xi_d,)$ and, $\boldsymbol{\eta} \equiv (\eta_1, \dots, \eta_h)$, by the relation

---

4 The equation of Kinetic Theory was proposed by Botzmann in 1872 when Markov was sixteen years old.

5 This assumption can be seen as equivalent to assuming a form of the Principle of Inertia for many-particle systems [12].

6 This simply means that the external events may be always described by a set of measures or numerical labels. And, provided we adopt a widely shared choice of a numerical labelling method, this is, always possible.

DOI: http://dx.doi.org/10.5772/intechopen.1004791

$$\begin{aligned} \rho_{\mathbf{E}_t}(\boldsymbol{\varepsilon}) &= \delta(\xi_1 - u_1(t))\,\delta(\xi_2 - u_2(t))\dots\delta(\xi_d - u_d(t))\times \\ &\times\delta(\eta_1 - v_1(t))\,\delta(\eta_2 - u_2(t))\dots\delta(\eta_h - v_h(t)) = , \qquad \forall t\in[0,T) \\ &= \delta(\boldsymbol{\xi} - \mathbf{u}(t))\,\delta(\boldsymbol{\eta} - \mathbf{v}(t)). \end{aligned} \tag{23}$$

This means that the process $\{\mathbf{E}_t\}_{t\in[0,T)}$ is a *deterministic* external process. For instance, we can imagine that our system consists of all the inhabitants of a given region, whose states are vectors, one component of which is the amount of money of each individual, and the other components are quantities of food. We consider only interactions consisting in purchase and sale of apricots. In the late summer, in the autumn, and in the winter the interaction rate will be zero (no apricots in these seasons); in the early springtime, the interaction rate will be small, and the transition probabilities will describe changes of states with little increases or decreases of the quantity of apricots, and large increases or decreases of the amount of money (in that period, apricots are hardly available and very expensive); in the late springtime and in the early summer, the interaction rate will attain its maximum value and the transition probabilities will describe changes of states with large increases or decreases of the quantity of apricots, and small increases or decreases of the amount of money (in that period, many apricots available at low prices). So, both the interaction rates and the transition probabilities depend on seasons, that is, are functions of time, and system (22) must be simply written in the form

$$\begin{aligned} \frac{p_{\boldsymbol{x},t}}{t}(\boldsymbol{x}_i) = \sum_{l=1}^{q}\Bigg\{ &\sum_{k\neq i}^{1\dots p} \tau_{kl}(t)p_{\boldsymbol{x},t}(\boldsymbol{x}_k)p_{\boldsymbol{y},t}(\boldsymbol{y}_l)P_{\mathbf{z}_{kl},\boldsymbol{x}_i\,|\,\chi_t^{kl}=1}(t) + \\ &-p_{\boldsymbol{x},t}(\boldsymbol{x}_i)\sum_{k\neq i}^{1\dots p} \tau_{il}(t)p_{\boldsymbol{y},t}(\boldsymbol{y}_l)P_{\mathbf{z}_{il},\boldsymbol{x}_k\,|\,\chi_t^{il}=1}(t)\Bigg\}, \qquad (i = 1, 2, \dots, l^s). \end{aligned} \tag{24}$$

Accordingly, the introduction of a dependence on time of both interaction rates and transition probabilities seems to be the most appropriate way to link the behavior of a many-particle system to deterministic external influences. Of course, the interesting aspect, and also the most difficult one, of this kind of model, is the choice of functions $\tau_{kl}(t)$ and $P_{\mathbf{z}_{kl},\boldsymbol{x}_i\,|\,\chi_t^{kl}=1}(t)$ in such a way as to express faithfully the changes of the environment, as well as the way in which these changes are perceived by the individuals of the system. At an initial stage, we can adopt a purely hypothetical method and make any assumption we want, but when we have in mind some precise applications, then we need a preliminary statistical analysis to get the right hints toward the most plausible assignment of the above functions.

7. Equations for many-particle systems with stochastic transition matrices

The case treated in the previous section is rather trivial, and its main interest lies in the choice of the dependence of all the interaction rates $\tau_{kl}(t)$ and of all transition probabilities $P_{\mathbf{z}_{kl},\boldsymbol{x}_i\,|\,\chi_t^{kl}=1}(t)$ on time, which is connected with the application we have in mind, that is, much more with experimental and statistical analysis than with theoretical reasonings. Much more interesting is the case in which the changes of external world influencing the behavior of the system are not assigned at each time, but *stochastic* events. This means that we have to consider again a stochastic process $\{(\mathbf{X}_t, \chi_t, \mathbf{E}_t)\}_{t\in[0,T)}$, but now, for any $t\in[0,T)$, $\mathbf{E}_t$ is a random variable in the "strict" sense of the word. More precisely, for any $t\in[0,T)$,

1. We denote again by $\mathbb{W}\equiv\mathbb{R}^d\times\mathbb{Z}^h$ the range of all possible values of $\mathbf{E}_t$.

2. We denote by $\boldsymbol{\varepsilon}\equiv(\boldsymbol{\xi},\boldsymbol{\eta})\equiv(\xi_1,\ldots,\xi_d,\eta_1,\ldots,\eta_h)$ any unspecified element of $\mathbb{W}$.

3. We set $\boldsymbol{a}\equiv(a_1,a_2,\ldots,a_d)$, $\boldsymbol{b}\equiv(b_1,b_2,\ldots,b_d)$, and $I(\boldsymbol{a},\boldsymbol{b})\equiv[a_1,b_1]\times[a_2,b_2]\times\ldots\times[a_d,b_d]$;

4. We set $\mathbf{E}_t\equiv(\mathbf{E}_{t,1},\mathbf{E}_{t,2})$ where the range of values of $\mathbf{E}_1$ is $\mathbb{R}^d$ and the range of values of $\mathbf{E}_2$ is $\mathbb{Z}^h$, and assume that $\mathbf{E}_{t,1}$ and $\mathbf{E}_{t,2}$ are independent;

5. To the (continuous) random variable $\mathbf{E}_1$ is associated a probability density function $\rho_{t,1}(\boldsymbol{\xi})\equiv\rho_1(\boldsymbol{\xi},t)$ such that

$$\mathbf{P}(\mathbf{E}_{t,1}\in I(\boldsymbol{a},\boldsymbol{b}))=\int_{I(\boldsymbol{a},\boldsymbol{b})}\rho_1(\boldsymbol{\xi},t)\,\mathrm{d}\boldsymbol{\xi}<1$$

for any $(\boldsymbol{a},\boldsymbol{b})\in\mathbb{R}^d\times\mathbb{R}^d$ such that $\mathbb{R}^d\backslash I(\boldsymbol{a},\boldsymbol{b})\neq\varnothing$;

6. To the (discrete) random variable $\mathbf{E}_2$ is associated a probability distribution function $\mathbf{P}_{t,2}$ such that

$$\mathbf{P}(\mathbf{E}_{t,2}=\boldsymbol{\eta})=\mathbf{P}_{t,2}(\boldsymbol{\eta})\equiv\mathbf{P}_2(\boldsymbol{\eta},t).$$

For any $\boldsymbol{\eta}\in\mathbb{Z}^h$, $\mathbf{P}_2(\boldsymbol{\eta},t)<1$.

7. we denote by the symbol $\tau_{ij}(\boldsymbol{\varepsilon},t)$ the density (on the interval $[0,T)$) of the probability that $\chi_t^{ij}=1$ conditional to the event $\mathbf{E}_t=\boldsymbol{\varepsilon}$.

Accordingly, for any $\boldsymbol{\varepsilon}\in\mathbb{W}$, the associated probability density of $\mathbf{E}_t$ at $\boldsymbol{\varepsilon}$ is

$$\rho_{\mathbf{E}_t}(\boldsymbol{\varepsilon})=\rho_1(\boldsymbol{\xi},t)\mathbf{P}_2(\boldsymbol{\eta},t).$$

This stated, according to the law of alternatives, and under the conditions imposed in Section 1.5, we find that each transition probability depends on time, and we have

$$P_{\mathbf{z}_{il},\boldsymbol{x}_k\,|\,\chi_{il,t}=1,t}=\sum_{\boldsymbol{\eta}\in\mathbb{Z}}\mathbf{P}_2(\boldsymbol{\eta},t)\int_{\mathbb{R}^d}\mathbf{P}\big(\mathbf{z}_{il},\boldsymbol{x}_k\,|\,\chi_{il,t}=1,\mathbf{E}_1=\boldsymbol{\xi},\mathbf{E}_2=\boldsymbol{\eta}\big)\rho_1(\boldsymbol{\xi},t)\tau_{ij}(\boldsymbol{\varepsilon},t)\,\mathrm{d}\boldsymbol{\xi}. \tag{25}$$

Accordingly, the system of equations governing the evolution of the many-particle system we have decided to describe by means of the present language and of the notions of random processes and, in particular, of Markov Chains, takes the form

$$\frac{\mathrm{d}p_{x,t}}{t}(\boldsymbol{x}_i) = \sum_{l=1}^{q} \sum_{k \neq i}^{1\ldots p} \sum_{\eta \in \mathbb{Z}} \mathbf{P}_2(\boldsymbol{\eta}, t) \times \left\{ p_{x,t}(\boldsymbol{x}_k) p_{y,t}(\boldsymbol{y}_l) \int_{\mathbb{R}^d} \mathbf{P}_{\mathbf{z}_{kl}, \boldsymbol{x}_i \mid \chi_t^{kl}=1, \mathbf{E}_t=\boldsymbol{\varepsilon}} \ \rho_1(\boldsymbol{\xi}, t)\, \tau_{kl}(\boldsymbol{\varepsilon}, t)\, \mathrm{d}\boldsymbol{\xi} + \right.$$

$$\left. - p_{x,t}(\boldsymbol{x}_i) p_{y,t}(\boldsymbol{y}_l) \int_{\mathbb{R}^d} \mathbf{P}_{\mathbf{z}_{il}, \boldsymbol{x}_k \mid \chi_t^{il}=1, \mathbf{E}_t=\boldsymbol{\varepsilon}} \ \rho_1(\boldsymbol{\xi}, t)\, \tau_{il}(\boldsymbol{\varepsilon}, t)\, \mathrm{d}\boldsymbol{\xi} \right\}, \qquad (i = 1, 2, \ldots, l^s) \tag{26}$$

where $\mathbf{P}_{\mathbf{z}_{kl}, \boldsymbol{x}_i \mid \chi_t^{kl}=1, \mathbf{E}_t=\boldsymbol{\varepsilon}} \equiv \mathbf{P}\left(\mathbf{z}_{kl}, \boldsymbol{x}_i \mid \chi_t^{kl} = 1, \mathbf{E}_1 = \boldsymbol{\xi}, \mathbf{E}_2 = \boldsymbol{\eta}\right)$ and $\mathbf{P}_{\mathbf{z}_{il}, \boldsymbol{x}_k \mid \chi_t^{il}=1, \mathbf{E}_t=\boldsymbol{\varepsilon}} \equiv \mathbf{P}\left(\mathbf{z}_{il}, \boldsymbol{x}_k \mid \chi_t^{il} = 1, \mathbf{E}_1 = \boldsymbol{\xi}, \mathbf{E}_2 = \boldsymbol{\eta}\right)$, when Ω (the state space common to all components of the vector Markov Chain X) is a discrete set (as we have assumed throughout the whole chapter till now). When Ω is instead a continuous set (typically, a real interval), then the sums in relations (26) must be replaced by integrals and — with the obvious correspondences $\boldsymbol{x}_i \to \boldsymbol{x}$, $\boldsymbol{x}_k \to \boldsymbol{x}'$, $\boldsymbol{y}_l \to \boldsymbol{y}$, $\mathbf{z}_{kl} \to (\boldsymbol{x}', \boldsymbol{y})$, $\mathbf{z}_{il} \to (\boldsymbol{x}, \boldsymbol{y})$, $\tau_{kl}(\boldsymbol{\varepsilon}, t) \to \tau(\boldsymbol{x}', \boldsymbol{y}; \boldsymbol{\varepsilon}, t)$, $\tau_{il}(\boldsymbol{\varepsilon}, t) \to \tau(\boldsymbol{x}, \boldsymbol{y}; \boldsymbol{\varepsilon}, t)$, $\chi_t^{kl} \to \chi(\boldsymbol{x}', \boldsymbol{y}, t)$, and $\chi_t^{il} \to \chi(\boldsymbol{x}, \boldsymbol{y}, t)$ — we get the equation[7]

$$\frac{\partial p_{\boldsymbol{x}}}{\partial t}(\boldsymbol{x}, t) = \sum_{\eta \in \mathbb{Z}} \mathbf{P}_2(\boldsymbol{\eta}, t) \times \int_{\Omega} \int_{\Omega}$$
$$\left\{ p_{\boldsymbol{x}}(\boldsymbol{x}', t) p_{\boldsymbol{y}}(\boldsymbol{y}, t) \int_{\mathbb{R}^d} \mathbf{P}_{(\mathbf{x}', \boldsymbol{y}), \boldsymbol{x} \mid \chi(\boldsymbol{x}', \boldsymbol{y}, t)=1, \mathbf{E}_t=\boldsymbol{\varepsilon}} \ \rho_1(\boldsymbol{\xi}, t) \tau(\boldsymbol{x}', \boldsymbol{y}; \boldsymbol{\varepsilon}, t)\, \mathrm{d}\boldsymbol{\xi} + \right.$$
$$\left. - p_{\boldsymbol{x}}(\boldsymbol{x}, t) p_{\boldsymbol{y}}(\boldsymbol{y}, t) \int_{\mathbb{R}^d} \mathbf{P}_{(\mathbf{x}, \boldsymbol{y}), \boldsymbol{x}' \mid \chi(\boldsymbol{x}, \boldsymbol{y}, t)=1, \mathbf{E}_t=\boldsymbol{\varepsilon}} \ \rho_1(\boldsymbol{\xi}, t) \tau(\boldsymbol{x}, \boldsymbol{y}); \boldsymbol{\varepsilon}, t)\, \mathrm{d}\boldsymbol{\xi} \right\} \mathrm{d}\boldsymbol{x}' \boldsymbol{y} \tag{27}$$

where we have also agreed to write

$$p_{\boldsymbol{x},t}(\boldsymbol{x}) = p_{\boldsymbol{x}}(\boldsymbol{x}, t), \qquad p_{\boldsymbol{x},t}(\boldsymbol{x}') = p_{\boldsymbol{x}}(\boldsymbol{x}', t), \qquad p_{\boldsymbol{y},t}(\boldsymbol{y}) = p_{\boldsymbol{y}}(\boldsymbol{y}, t).$$

Equation (27) is a generalization of system (24). As a matter of fact, this latter can be easily obtained from the former by simply replacing the probability density function ρ_1 with the one defined by relation (23). And, of course, the equations governing the evolutions of *isolated* many-particle systems can be derived at once by choosing $\{\mathbf{E}_t\}$ as a *deterministic* and *constant* process (i.e. a process in which all the random variables $\mathbf{E}_t$ have the same probability density function

$$\rho(\boldsymbol{\xi}) = \delta(\boldsymbol{\xi} - \mathbf{e})$$

on $\mathbb{W}$, however, $\mathbf{e}$ is fixed). So, system (27) seems to be the most general description of the behavior of any one of such systems. Recently, however, other types of external influence have been considered, acting directly on the state vectors and not on the interactions between particles [13, 14]. This point will be discussed in some details in the last section of the chapter.

---

[7] Notice that system (24) is replaced by only one equation because the l^s unknown functions $p_{x,t}(\boldsymbol{x}_i)$ of one variable (the time variable) corresponding to the l^s possible states are replaced, in view of the continuity of the state variable, by a unique unknown function of *both* the variables $\boldsymbol{x}$ and t.

8. Subsystems

This section is almost a pure aside, just to give a complete exposition of the development of the whole theory of many-particle systems as based on the theory of stochastic processes. We want to recall that a more general formulation of the theory takes into consideration the case in which a system $\boldsymbol{\mathcal{S}}$ is the join of a number N of subsystems $\boldsymbol{\mathcal{S}}_1, \boldsymbol{\mathcal{S}}_2, \dots, \boldsymbol{\mathcal{S}}_N$, called the *functional subsystems* of $\boldsymbol{\mathcal{S}}$, that can share the same state space or not, and are such that, for any $\lambda \in \{1, 2, \dots, N\}$, the particles of $\boldsymbol{\mathcal{S}}_\lambda$ are not constrained to share the same state. In order to show the way in which the equations describing the behavior of $\boldsymbol{\mathcal{S}}$ must be modified under this assumption, in order to avoid an undue multiplication of indexes, and for the sake of a better readability, we shall assume that all the subsystems share a continuous state space $\Omega \equiv (a, b) \subseteq \mathbb{R}$. The point of the present development is that now we must allow interactions between particles of different subsystems so that we must consider a mixed stochastic process $\{(\mathbf{X}_t, \boldsymbol{\chi}_t, \mathbf{E}_t)\}_{t \in [0,T)}$ where $\boldsymbol{\chi}_t \equiv (\chi_{\lambda,\mu}(\boldsymbol{x}, \boldsymbol{y}, t))$, and for any $t \in [0, T)$ and for any 4-tuple $(\lambda, \mu, \boldsymbol{x}, \boldsymbol{y}) \in \{1, 2, \dots, N\}^2 \times \Omega^{2s}$ the random variable $\chi_{\lambda,\mu}(\boldsymbol{x}, \boldsymbol{y})$ is again a Bernoulli variable, with probability density function $\tau_{\lambda\mu}(\boldsymbol{u}, \boldsymbol{v})\Delta t$, in view of the continuity of variables $\boldsymbol{x}$ and $\boldsymbol{y}$. By reproducing almost without changes the reasonings of the previous two sections, we obtain the system

$$
\begin{aligned}
\frac{\partial p^\lambda_{\boldsymbol{x}}}{\partial t}(\boldsymbol{x}, t) = &\sum_{\boldsymbol{\eta} \in \mathbb{Z}} \mathbf{P}_2(\boldsymbol{\eta}, t) \times \sum_{\mu=1}^{\nu} \int_\Omega \int_\Omega \\
&\Bigg\{, p^\lambda_{\boldsymbol{x}}(\boldsymbol{x}', t) p^\mu_{\boldsymbol{y}}(\boldsymbol{y}, t) \int_{\mathbb{R}^d} \mathbf{P}_{(\boldsymbol{x}', \boldsymbol{y}), \boldsymbol{x} \mid \chi_{\lambda\mu}(\boldsymbol{x}', \boldsymbol{y}, t) = 1, \mathbf{E}_t = \boldsymbol{\varepsilon}} \; \rho_1(\boldsymbol{\xi}, t) \tau_{\lambda\mu}(\boldsymbol{x}', \boldsymbol{y}; \boldsymbol{\varepsilon}, t) \, \mathrm{d}\boldsymbol{\xi} + \\
&\quad - p^\lambda_{\boldsymbol{x}}(\boldsymbol{x}, t) p^\mu_{\boldsymbol{y}}(\boldsymbol{y}, t) \int_{\mathbb{R}^d} \mathbf{P}_{(\mathbf{x}, \boldsymbol{y}), \boldsymbol{x}' \mid \chi_{\lambda,\mu}(\boldsymbol{x}, \boldsymbol{y}, t) = 1, \mathbf{E}_t = \boldsymbol{\varepsilon}} \; \rho_1(\boldsymbol{\xi}, t) \tau_{\lambda\mu}(\boldsymbol{x}, \boldsymbol{y}); \boldsymbol{\varepsilon}, t) \, \mathrm{d}\boldsymbol{\xi} \Bigg\} \\
&\quad \mathrm{d}\boldsymbol{x}' \boldsymbol{y} \quad (\lambda = 1, 2, \dots, N).
\end{aligned}
\tag{28}
$$

We have now again a system of equations. Each of them describes the evolution of the probability density function of a corresponding subsystem in consequence of the interaction of both interactions between particles of the same subsystem and of particles of different subsystems. But, of course, we have written the equations in the final form they have when also external actions that modify both the rates and the effects of interactions.

9. Alternative views, conclusions and perspectives

In almost all the literature about many-particle systems, external actions are completely disregarded, and the system of equations governing their evolution is written in its simplest form, where both the transition probabilities and the interaction rates are constant. Only recently, a number of papers have taken into account external «forcing» events that are capable to modify the distribution of states on each subsystem of a system $\boldsymbol{\mathcal{S}}$ independently of possible interactions between particles [11–15]. These external events, however, are assumed to act *directly* on the functions $p^\lambda_{\boldsymbol{x}}(\boldsymbol{x}, t)$ (or $p^\lambda_{\boldsymbol{x}}(\boldsymbol{x}, t)$, when the state space is discrete). The origin of this viewpoint is in [11] where, dealing with a description of possible evolutions of an epidemics, an additional term describing spontaneous healing or death was necessary to get a faithful picture. This also led to general considerations about the implicit use of an analog of the Principle of Inertia in the scheme depicted in Sections 1.5 and 1.6. Taking into account an additional term of this type, system (28) becomes

$$\frac{\partial p_{\boldsymbol{x}}^{\lambda}}{\partial t}(\boldsymbol{x},t) = \sum_{\eta \in \mathbb{Z}} \mathbf{P}_2(\boldsymbol{\eta},t) \times \sum_{\mu=1}^{\nu} \int_{\Omega} \int_{\Omega}$$

$$\Big\{, p_{\boldsymbol{x}}^{\lambda}(\boldsymbol{x}',t) p_{\boldsymbol{y}}^{\mu}(\boldsymbol{y},t) \int_{\mathbb{R}^d} \mathbf{P}_{(\boldsymbol{x}',\boldsymbol{y}),\boldsymbol{x} | \chi_{\lambda\mu}(\boldsymbol{x}',\boldsymbol{y},t)=1,\mathbf{E}_t=\boldsymbol{\varepsilon}} \; \rho_1(\boldsymbol{\xi},t) \tau_{\lambda\mu}(\boldsymbol{x}';\boldsymbol{y},\boldsymbol{\varepsilon}t) \; \mathrm{d}\boldsymbol{\xi} +$$

$$-p_{\boldsymbol{x}}^{\lambda}(\boldsymbol{x},t) p_{\boldsymbol{y}}^{\mu}(\boldsymbol{y},t) \int_{\mathbb{R}^d} \mathbf{P}_{(\mathbf{x},\boldsymbol{y}),\boldsymbol{x}' | \chi_{\lambda,\mu}(\boldsymbol{x},\boldsymbol{y},t)=1,\mathbf{E}_t=\boldsymbol{\varepsilon}} \; \rho_1(\boldsymbol{\xi},t) \tau_{\lambda\mu}(\boldsymbol{x},\boldsymbol{y}); \boldsymbol{\varepsilon},t) \; \mathrm{d}\boldsymbol{\xi}\} \Big\} \, \mathrm{d}\boldsymbol{x}' \boldsymbol{y} +$$

$$+F^{\lambda}\left[p_{\boldsymbol{x}}^1, p_{\boldsymbol{x}}^2, \dots, p_{\boldsymbol{x}}^N\right](t) \qquad (\lambda = 1, 2, \dots, N). \tag{29}$$

where the form of the term $F^{\lambda}\left[p_{\boldsymbol{x}}^1, p_{\boldsymbol{x}}^2, \dots, p_{\boldsymbol{x}}^N\right](t)$ depends on the context and on the type of external solicitation producing a «spontaneous» variation of the distributions.

But now, to conclude this overview of the possible developments of the modern generalization of kinetic theory, that becomes evident when it is analyzed and formalized in terms in terms of n-tuples of joint random processes (one of which is a Markov Chain, but to this point we will return at the very end of this final section), going back to the discrete version of the equation, from which we started to reconstruct the logic the equations of the theory are based upon, we want to point out an interesting new application of the theory in the scope of experimental procedures. This application was presented in [17] and its theoretical formulation in [18].

In order to outline the above-mentioned development, we first assume to deal with a system $\boldsymbol{\mathcal{S}}$ split into two subsystems $\boldsymbol{\mathcal{S}}_1 = \{\Sigma\}$ and $\boldsymbol{\mathcal{S}}_2 = \{X_h\}_{h \in \mathbb{N}}$. So, $\boldsymbol{\mathcal{S}}_1$ contains *only one* element, which will be called the «subject», while the elements of $\boldsymbol{\mathcal{S}}_2$, that are random variables[8]will be called the *experiments* or *questions* about Σ. The «state space of the subject Σ» is a set $\Omega_1 \equiv \{\boldsymbol{x}_1, \boldsymbol{x}_2, \dots, \boldsymbol{x}_p\}$ while, for any $h \in \mathbb{N}$, the «state space of question X_h», that is, the set of all of its possible outcomes, is the set $\Omega_{2h} \equiv \left\{y_{h1}, y_{h2}, \dots, y_{hq_h}\right\} \subset \mathbb{Z}$. The set $\Omega_2 = \cup_{h \in \mathbb{N}} \Omega_{2h}$ will be the state space common to all the random variables of the sequence $\{X_h\}_{h \in \mathbb{N}}$. We consider the discrete (stepwise) stochastic process $\{(\mathbf{X}_h, \chi_h)\}_{h \in \mathbb{N}}$, where $\{\mathbf{X}_h\}$, such as in Section 1.3, is a Markov Chain with the state space $\Omega_1 \times \Omega_2$ and

$$\boldsymbol{\chi}_h \equiv \begin{pmatrix} \chi_h^{11}(x_i, x_j) & \chi_h^{12}\left(x_i, y_{hj}\right) \\ \chi_h^{21}\left(x_i, y_{hj}\right) & \chi_h^{22}\left(y_{hi}, y_{hj}\right) \end{pmatrix}$$

for any $h \in \mathbb{N}$. The entries of this matrix are independent Bernoulli's random variables. $\chi_h^{11}(x_i, x_j) = 1$ means that an interaction between Σ in the state Σ and Σ in the

[8] The reader should carefully note that, differently from any classical application of the description of many-particle systems outlined in the previous Section of this chapter, the members of system $\boldsymbol{\mathcal{S}}$ can be very different from «particles» in the classical sense of the word, *that is*. abstract objects, such as random variables.

state x_j occurs at h-th step, and $\chi_h^{22}\left(y_{hi}, y_{hj}\right) = 1$ means that at the same step an interaction occurs between two different answers to the same question X_h. We have not yet defined the meaning of the word «interaction» in the present context (this point will be briefly discussed at the end of the section), so for the moment, we simply assume that $\chi_h^{11}(x_i, x_j) = \chi_h^{22}(y_{hl}, y_{hm}) = 0$ for any $h \in \mathbb{N}$ and for any $(i,j) \in \{11, 2, \ldots, p\}^2$ and $(l,m) \in \{1, 2, \ldots, q_h\}^2$. If $\tau_{i,j}^{11} = \mathbf{P}(\chi_h^{11}(x_i, x_j) = 1)$ and $\tau_{l,m}^{22} = \mathbf{P}\chi_h^{22}(y_{hl}, y_{hm}) = 1)$, this also means that $\tau_{i,j}^{11} = \tau_{l,m}^{22} = 0$. On the other hand, $\chi_h^{12}\left(x_i, y_{hj}\right) = 1$ and $\chi_h^{21}\left(x_i, y_{hj}\right) = 1$ mean that an interaction between Σ in the state x_i and the question X_h giving the answer y_{hj} certainly occurs at h-th step; this corresponds to the fact that the interaction is decided and produced by an experimenter. We assume $\chi_h^{12}\left(x_i, y_{hj}\right) = \chi_h^{21}\left(x_i, y_{hj}\right) = 1$ for any $h \in \mathbb{N}$ and for any $i \in \{11, 2, \ldots, p\}$ so that $\tau_{i,j}^{12} = \tau_{i,j}^{21} = 1$. Moreover, we assume $P^{21}_{\left(x_k, y_l\right), x_i | \chi_h^{21}\left(x_i, y_{hj}\right) = 1} = 0$ for any $h \in \mathbb{N}$ and for any $(i,j) \in \{11, 2, \ldots, p\}^2$. All the other conditions assuring that we are allowed to write equations (19) are assumed to hold, These equations, with obvious meaning of the symbols, will be finally rewritten in the form

$$
\begin{aligned}
& p^1_{x,h+1}(x_i) - p^1_{x,h}(x_i) = \\
& \quad = \sum_{l=1}^{q_h} \left\{ \sum_{k \neq i}^{1\ldots p} p^1_{x,h}(x_k) p^2_{y,h}(y_{hl}) P^{12}_{\left(x_k, y_{hl}\right), x_i | \chi^{12}\left(x_i, y_{hi}\right) = 1} + \right. \\
& \quad \left. - p^1_{x,h}(x_i) \sum_{k \neq i}^{1\ldots p} p^2_{y,h}(y_{hl}) P^{12}_{\left(x_i, y_{hl}\right), x_k | \chi^{12}\left(x_i, y_{hi}\right) = 1 = 1} \right\}, \qquad (i = 1, 2, \ldots, p). \quad (30) \\
& p^2_{y,h+1}\left(y_{(h+1)i}\right) - p^2_{y,h}(y_{hi}) = 0
\end{aligned}
$$

To conclude the section and the chapter, we want now to expose an interpretation as complete as possible of the special model just described. To this aim, what is necessary to stress first is that system (30) is meant to describe the results of interactions between *information* rather than between objects. More precisely, the values x_i of the variable in Ω_1 should be interpreted as labels that *define*, in the appropriate context, possible states of the subject; then, the interaction between the subject and the h-th experience X_h on it places this latter into a *state*, expressed by a real value y_{hi}, that is, the outcome of a measurement of a physical parameter. Now, in virtue of the information carried by the value y_{hi}, the i-th experience *interacts* with the initial definition, and in the sense that it can either confirm the initial state x_i (by increasing its probability) or raise doubts about it (by lowering its probability). Accordingly, the terms $p^1_{x,h}(x_i)$ and $p^1_{x,h}(x_i)$ in equations (30) must be interpreted respectively as the *absolute* probability that the *definition* of the state of the subject before the experience is the value x_i and the *absolute* probability that the outcome of the h-th experience is the state y_{hi}; furthermore, the term $\boldsymbol{P}^{12}_{\left(x_k, y_{hl}\right), x_i | \chi^{12}\left(x_i, y_{hi}\right) = 1}$ is the probability that definition x_k, after the «interaction» with information y_{hl} *becomes* definition x_i.

In [17], equations (30) were derived by discretization from the ones of kinetic theory and used to describe the way in which new experiences on a number of different parameters can modify the classification of a phenomenon, in particular the diagnosis of depression in a human subject suspected to be depressed according to the score obtained by answering a questionnaire. Of course, in an application like that, the transition probabilities must be estimated in advance by a preliminary study of the correlation of the involved random variable with the possible depression states. We have reported this application, together with the classical application of kinetic theory and its developments, to give a more complete picture of the way in which the study of joint random processes (vector random processes), in particular when one of the processes considered simultaneously is a Markov Chain, can open the way to an ever wider range of applications.

Author details

Bruno Carbonaro* and Federica Vitale
Dipartimento di Matematica e Fisica, Università degli Studi della Campania "L. Vanvitelli", Caserta, Italy

*Address all correspondence to: bruno.carbonaro@unicampania.it

References

[1] Gilch L. Markov Chains: An Introduction: Lecture Notes. Independently Published. 2022

[2] Lawler GF. Introduction to Stochastic Processes. Boca Raton: Chapman and Hall/CRC; 2006

[3] Norris JR. Markov Chains. Cambridge: Cambridge University Press; 1998

[4] Boltzmann L. Lectures on Gas Theory. Courier Corporation; 2012

[5] Aylaj B, Bellomo N, Gibelli L. Crowd Dynamics by Kinetic Theory Modeling: Complexity, Modeling, Simulations, and Safety. New York: Springer; 2020

[6] Bellomo N, Bellouquid A, Gibelli L, Outada N. A Quest towards a Mathematical Theory of Living Systems. Basel: Birkhäuser; 2017

[7] Bellomo N, Brezzi F. Traffic, crowds and swarms. Mathematical Models and Methods in Applied Sciences. 2008;**18** (Suppl):1145-1148

[8] Bellomo N, Degond P, Tadmor E, editors. Active Particles, Volume 1: Advances in Theory, Models, and Applications. Basel: Birkhäuser; 2017

[9] Bertotti ML, Modanese G. Economic inequality and mobility in kinetic models for social sciences. The European Physical Journal Special Topics. 2019; **225**(10):1945-1958

[10] Chinesta F, Abisset-Chavanne E. A Journey Around the Different Scales Involved in the Description of Matter and Complex Systems: A Brief Overview with Special Emphasis on Kinetic Theory Approaches. New York: Springer; 2017

[11] Carbonaro B. Modeling epidemics by means of the stochastic description of complex systems. Computational and Mathematical Methods. 2021;**3**(6): 1208-1220

[12] Carbonaro B. The role of the principle of inertia in KTAP models. Journal of Mathematical Physics. 2022; **63**(1)

[13] Carbonaro B, Menale M. A nonconservative kinetic framework under the action of an external force field: Theoretical results with application inspired to ecology. European Journal of Applied Mathematics. 2023;**34**(6): 1170-1186

[14] Carbonaro B, Menale M. The behavior of a many-particle system with fading memory interacting with external world, to appear

[15] Menale M, Munafò CF. A kinetic framework under the action of an external force field: Analysis and application in epidemiology. Chaos, Solitons & Fractals. 2023;**174**:113801

[16] Benaim M, Hurth T. Markov Chains on Metric Spaces: A Short Course. New York: Springer Nature, Springer; 2022

[17] Vitale F, Carbonaro B, Esposito A. A dynamic probabilistic model for heterogeneous data fusion: A Pilot Case study from computer-aided detection of depression. Brain Science. 2023;**13**: 1339-1359

[18] Vitale F, Carbonaro B, Esposito A. A methodological application of a time-discrete version of the kinetic-theoretic language to interactions between random variables. To appear

Chapter 5

Brownian Computing and Its Similarities with Our Brains

Alexandra Pinto

Abstract

With the current need for the miniaturization of electronic circuits and transistors, there is a critical problem of fluctuations that create signal interference for the outputs of those circuits. One solution is to use those fluctuations and stochastic manipulation to the benefit of computing architectures. Such a solution is defined as Brownian computing, and it is possible to apply it with Brownian circuits. At those physical limits, nonlinear dynamics dominate, and synaptic modeling is critical. Synapses are the target of most, if not all, brain disorders. Because of the enormous number of synapses in our brain (~ 100,000,000,000), it is often difficult to comprehend how defects in one type of synapse yield the overall loss in brain performance characteristic of all brain disorders. Computational modeling of synapses has been a slow process that is constrained by advances in the interdisciplinary fusion of physics, both for measuring and data analysis, and advances in molecular biology or biophysics. These constraints are also computational: some models have detailed sets of coupled equations with experimental parameters that, although accurate, are impossible to solve because of the enormous number of synapses. This limitation impedes understanding of brain information processing, both in health and in disease states.

Keywords: Brownian computing, beyond von-neumann computing, synapses, memristors, non-linear dynamics

1. Introduction

With the current need for miniaturization of electronic circuits and transistors, there is a critical problem of fluctuations that create signal interference for the outputs of those circuits. One solution is to use those fluctuations and stochastic manipulation to the benefit of computing architectures. Such a solution is defined as Brownian computing, and it is possible to apply it with Brownian circuits [1]. At those physical limits, nonlinear dynamics dominate.

In this chapter, we are going to focus on one of the similarities between our brain and one type of brain-inspired computing approach known as reservoir computing (RC) [2], which uses to its benefit the nonlinear dynamics of an echo state to map a complex problem to a much simpler linear problem, a well-know concept in computer science - problem reduction. Our brains are constantly transforming our inputs into a simpler representation through the use of nonlinear dynamics to obtain perception and action in a synchronous way. In the next section, I will introduce an analogous source of fluctuations in our brains compared to the miniaturization of transistors. We will focus on synaptic modeling as they are the main source of nonlinearity and fluctuation in the brain, and they will play the role of echo state reservoir in analogy with reservoir computing. Tunning synaptic weight during learning is equivalent to biasing the operating point with the corresponding slope of the flux versus charge dynamics, where the synaptic strength is equivalent to the memristance [3].

2. Computation with fluctuations in the brain: a quantized approach

Networks of neurons have the task of conveying a digital n-dimensional input space, which is comprised of each signal detected by each neuron, into an analogous four-dimensional space, that is our output space, the environment we perceive dynamically as a time-space continuum. This transformation from input to output must be created both in a continuous manner according to classical mechanics and conditioned to the Plank constant, the minimum action according to quantum theories, such that our perception of time and space could be smooth and efficient so as to minimize the errors in perception.

The brain uses these mechanisms for the projection of what we perceive as reality. This can only happen by computing with noise or spontaneous fluctuations of periodic origin in chemical synapses [4]. This is analogous to Brownian circuits and this synaptic modeling with memristors mechanisms, and what could be seen as a mesoscopic scale phase transition prediction by considering long-range correlations through the Bose-Einstein condensate. This can only be done with memristors since the flux-to-charge relation captures the mesoscopic and quantum effects needed.

As a side effect of the model, it gives us the Fokker-Einstein equation as a result of dynamical considerations. The scales and frequencies between the study of Brownian circuit fluctuation, lasers, and the synaptic cleft are at the limit between classical and quantum effects. This is why we can understand this model as a memristor and it is the generation of synaptic self-sustained oscillations. It is relevant to explore its connection with fluctuations of Quantum origin.

The before mentioned proposition implies that synaptic behavior cannot only be added by the Hebbian rule, but it has to be added as oscillations accounting for the Quantum effects that the dispersion pattern of interference creates. As a consequence of the Quantum correction for the interference in neuronal activity, we can understand the inter-neural communication as the particle-wave duality principle in neurons.

This proposition is necessary because input information in the nervous system is in the form of an analogous continuous wave that guarantees the flow and appropriate match of information. Since there is a change in the biophysical and chemical substrates from pre-synapse hydrophobic composition and its electromagnetic content to the more open hydrated space of the synaptic cleft, which affects the biophysical and electromagnetic characteristics of the post-synapse.

The nature of information along the transmission must carry a transformation from the wave state of the pre-synapse to the particle probabilistic state of the post-synapse. This transformation is carry out at the synaptic cleft, where, as a consequence of the diffracted interference pattern, the wave probability density is calculated in the Quantum Mechanical formalism. The most relevant characteristic of this model can be seen in learning and subsequently in memory, given that the oscillatory state at the synapse has a recurrent update given by the action potential probabilistic state that depends on the postsynaptic effect. This behavior captures attenuation which is the basis for learning.

There is a big impact on this proposition for memory and plasticity given that only recurrent patterns are constructing a stable channel at the post-synapse. However, its impact is not necessarily local given the effect of the quantum mechanical considerations; non-local quantum effects can be carried out by long-range correlations on the postsynaptic probability pattern. For this special consideration, we need to go beyond quantum effects and add fluctuating effects.

Even though the origin of the effect of complex, it is possible to simplify the model to a deterministic point. This does not imply that decision-making is fully determined, but the neural activity, as a physical substrate, can be fully determined by energetic considerations. Energy quantization can be used in a deterministic fashion for prediction purposes.

3. Synapses, source of fluctuations in the brain

Fluctuations in the brain can be modeled as a wave-to-pulse-to-wave conversion in the neuron's trigger zones, associated with the chemical synapses occurring in the nervous system. The wave-to-pulse-to-wave conversion was deeply studied by Freeman in the mesoscopic scale [5]. The current experimental database and theoretical explanation on this subject are very limited on the explanation of the different process that convert electrical into chemical and back to electrical energy via the synaptic cleft, where conservation of information and frequency must happen [4]. The interest in studying this problem arose from the fact and knowledge that the trains of action potentials certainly play a part in transmitting information through the nervous system, together with the fact that most of the time neurons are silent, without transmitting action potentials. Generally speaking, it is observed that 99 percent of the time, the electrical activity of neurons is in the subthreshold regime [6].

It is the combination of those low-amplitude fluctuating signals, without the energy to emanate an action potential, carrying the majority of information, that after broadcasting and at higher cortical levels of integration, generate the necessary knowledge and meaning that allows us to: a) perceive the world subjectively, b) communicate via the use of language, c) retain valuable memories, and d) in general, perform activities that do not require the fast response inter neuron communication that is associated with electric synapses.

The model that is proposed in this chapter is of a synapse as an oscillator. This oscillator undergoes constructive and destructive interference patterns of wave activity. The function of this wave patter is to smooth the train of action potential *via* the release of neurotransmitters from the presynaptic neuron into the post-synaptic neuron. The frequency is preserved during the process of transforming electrical into chemical and back to electrical energy. In a summary, in this theoretical proposal, synapses are described as a system composed of an input wave that is transformed quantum mechanically through interferometry.

A synapse is replaced by an oscillation that depends on the wavelength of the input signal. The collective synaptic interference pattern of waves will determine the points of maximum amplitude for the density wave synaptic function, where the action potential is observed.

There is a relation between the pulse train arriving at the post-synaptic neuron terminal and the release of neurotransmitters in order to manifest a chemical wave-like pattern that finally is converted, in the post-synaptic neuron, to a smoother electrical wave-like pattern and hence the term pulse wave conversion. Note that this model also describes a wave chemical to wave electrical conversion, contrary to the current assumption in neuroscience. This new proposal for the synapse as a memristor [7], allows us to recover the density function of the postsynaptic neuron. The final electrical activity resting potentials, described as a wave pattern, will, when above threshold, generate a new action potential and hence the term wave-to-pulse conversion [8]. Memristors also capture the memory capabilities of synapses and the nonlinearity chaotic behavior of the action potential.

4. Why oscillations represented by memristors at synapses?

Memristors, as well as synapses, store information, and the continuous flow of information needs a flux charge measure. The slope of such a relation is measured by the variable amount of neurotransmitter at the synaptic cleft. This is why we model the synapse as an oscillator; besides, oscillations in nature are ubiquitous (mechanical (movement of fluids)) and electromagnetic (movement of ions).

Most of the time, neurons are not evoking action potentials, but they are not at the zero equilibrium point either, rest state requires the continuous flow of information based mainly on fluctuations at synapses. Transformation from the synaptic fluctuating nonlinear state to the action potential state requires the addition and subtraction of memristors, forming the interference pattern that depends on other synapses connected to the same neuron, which as a result forms the wave probability density.

The model of the synapse is the same for all of them (biologically plausible oscillator), but the location is special in the influence of the pattern formation (not always constructively). Learning in the synapse (wave state) updates depending on the action potential (particle state). As the uncertainty principle, where the certainty of position is constrained by certainty in momentum and vice-versa.

Constant oscillation at the interior of the membrane; the potential is never zero, there is a non-stopping flow of information that happens without accounting for action potential initiation, but assuming that it is interference is also responsible for background noise generation, then is a cyclic process that is sensible to small perturbations. Evoking an action potential can be observed as the fact that the wave synaptic state functions overlap and cause the locking of their phases, creating a stronger state.

Given that spontaneous oscillations are an important assumption, considering the most common scenario in the lifespan of a neuron where it is 99 percent of the time in a sub-threshold regime and many of them die without ever making a spike, we do not need a forcing term in the oscillator modeling, but rather a very small initial condition that should move the system from the unstable origin. Neurons, in the absence of action potential stimulation, are in a latent constant transmission of information in the form of small amplitude waves that reside in the sub-threshold regime and which spontaneously emerge as a consequence of constructive interference. In this case, the resonant state is found for slightly perturbations of the initial condition.

DOI: http://dx.doi.org/10.5772/intechopen.1005437

Modeling synapses has been a slow process that is constrained by advances in the interdisciplinary fusion of physics, both for measuring and data analysis, and advances in molecular biology or biophysics [9]. Nonetheless, nowadays, the constraints are also computational, given that some models have detailed sets of coupled equations with experimental parameters that are pretty accurate but solving them for each one of the 5×1013 synapses is an impossible computational task with current computational resources. The main con-sequence of this computational limit is that there are different forms of synapse modeling, from detailed biophysical models that sacrifice the possibility of analyzing the behavior of several synapses to reductionist models that consider the phenomenon in a probabilistic way. Analyzing synapses with the wave-particle viewpoint is computationally efficient [10].

5. Electronic modeling of synapses

In this section, we will consider the electronic modeling of synapses with memristors. We will also add the proposed synaptic interference for computing collective behavior as synchronous oscillators. The memristor act as a synapse with the energy source through a nonlinear resistor with pumping effect that is connected with an inductance for magnetic field connectivity constraint. There is generation of action potential at locations of maximum amplitude at memristors interference pattern. Analog with double slit problem is for wave interference representation and waveparticle duality or synaptic-action potential dynamics. Synapses act as memristors that generate waves interfering to create a probability density of action potential generation in the form of constructive interference. This means that synapses cannot be added as discrete particles but they need to be added by the alternative approach of constructive and destructive interference.

Synaptic activity as a source of fluctuations for Brownian computing with interference is valuable for reducing computational costs. There are possible tunneling effects at resonance points since we are interested in the subthreshold regime where the amplitude of the signal is very small in comparison with the size of the barrier potential. Computation with uncertainty is possible. Both inhibitory and excitatory synapses (consequently excitatory and inhibitory neurons) can be modeled with the same model, memristors, which makes the computation easier than under considerations of ionic activity; however, inhibition is expressed as destructive interference and excitatory with constructive interference.

6. Algorithmic explanation

I. Compute the neural transmission as a function of the distribution of the post-synaptic neurons from the delta distribution of observations: this means the distribution of features, depending on k, the sub-network within each node in the network, and the delta distribution.

II. The weight matrix connects pre-synaptic neurons with post-synaptic neurons. In this case, the weight matrix is the synapse model of diffusion determined by oscillators with memristors elements. The weight matrix depends on the generative function of the post-synaptic neurons; that means the current prediction as a function of the neurons positions.

III. Calculate the synaptic weight for each particle, using each calculated synaptic weight matrix to update the state of the pre-synaptic and post-synaptic neuron for each neuron.

IV. The update of the pre-synaptic neuron between the distribution of observations and the generative function of the current prediction. In this case, it is the communication between pre-synaptic and post-synaptic neuron that will change the current state of the pre-synaptic neuron through interference oscillatory patterns that reflect the duality particle-wave properties.

V. Given that the communication between pre- and post-synaptic neurons is in both directions, the post-synaptic neurons require an update rule that depends on the pre-synaptic. Pre-synaptic neurons are the input that transmits sensory information; that means that even if communication happens with a preferential directional stream, it does not mean that in the opposite direction, there could not be a response effect that reflects the action-reaction nature of states in the particle-wave nature of communication. This update is, in this case, given by addition of the current post-synaptic neuron state, the deterministic classical drift function, the synaptic connectivity from the distribution of pre-synaptic neurons, and lastly, a Brownian motion term that comes from the fact that the process requires a spontaneous component.

VI. The algorithm finishes with the total number of neurons.

VII. At the end, the distribution of post-synaptic neurons or in this term, the wave distribution of the postsynaptic neuron is found after each neuron passes through the weight matrix or in this term the splitting phenomenon produced by the diffusion process that waves passing through the synaptic cleft have to experience. Once these waves get in contact with the wall of postsynaptic activity distribution, there is a reflective component that changes the activity of the pre-synaptic distribution.

7. Mathematical foundation

Analyzing synapses with the wave-particle viewpoint is computationally efficient for studies of memory and learning, where it is crucial to detect the appearance and disappearance of synaptic contacts and plasticity. With this framework, synapses are amplitudes of wave activity that vanish or emanate depending on the collective oscillatory behavior. This is a novel way for modeling interneuron communication that does not reside on the complexity of detailed biophysical neuronal models or its synapses, whereas is presented in this document, it is necessary to compute around 25 coupled differential equations.

Another desirable characteristic of this proposal is the fact that the required noise generation, proper of real synaptic recordings, is implicit in the wave pattern generation as a consequence of the distribution and structural complexity that can be reduced as the location of the slit, again as analogous to Young's experiment. This is a valuable feature, given that it is not necessary to add extra noise terms that, in the case of biophysical models or integrate and fire models, make the system harder to solve computationally.

This study is a potential tool for the analysis data where it is required a transformation from the postsynaptic signal to the presynaptic or vice-versa. This approach can also be useful in computational tasks, both in software and hardware, that can make use of real biological network architectures as templates of interference generators that are able to solve different computations.

Simplified explanation of the model of synaptic wave signal detection: Suppose that the distance and speed of a synaptic signal are to be determined. For this purpose, a transmitter/receiver neuron sends a test signal action potential signal f, toward the synaptic wave. Then, the signal f is reflected by the synaptic wave, and a part of the reflected signal is received by the neuron as an echo. The sent test signal is a pulse of short duration:

$$f(x) = \sigma(x)\exp(2\pi i\omega \cdot x) \tag{1}$$

with an envelope signal σ of slow variation. This means that supp σ transform is contained between -A and A and A is small compared to the carrier frequency ω. The support of f transformed is contained between $\omega - A$ and $\omega + A$.

Considering r as the distance between pre and postsynaptic neurons the transmitter and receiver, with v as the relative velocity between the receiver and the synaptic signal and with c as the speed of light. Then the "echo" (in analogy to the echo state in reservoir computing which is also represented by synaptic activity) is received with a time lag

$$\delta t = \frac{2r}{c} \tag{2}$$

Each frequency, f undergoes a Doppler shift $\delta\omega = -\frac{2\omega v}{c}$. Since the bandwidth 2A is small compared to the carrier frequency ω, the frequency shifts can be approximated by the shift $\delta\omega \approx -\frac{2\omega v}{c}$, independently of the exact form of f. No further distortion of f is taken into account, therefore the "echo" has the form:

$$e = M_{\delta\omega} T_{\delta t} f \tag{3}$$

At the receiver neuron the "echo" is compared to the time-frequency shifts of the original signal f by taking the correlation absolute value of the inner product of the "echo" signal with the $M_\omega T_x f$. Which is equivalent to the absolute value of the Short-time Fourier transform (STFT) of $f(x) = -\delta$ and $\omega - \delta\omega$ which is equal to the ambiguity function of the same function.

The values of δt and δ_ω and the distance r and the velocity v of the synaptic wave, can be determined by means of the lemma: Suppose that f is part of the Hilbert space L2 of Rd, and f is different from zero, then the absolute value of the ambiguity function is less than the ambiguity function of f evaluated at the Origen. Which is equal to the $\|x\| \forall \omega$ and x different from 0.

Synaptic wave is the support of the function σ with which the Wigner distribution should be convoluted in order to average each of its points. The lack of positivity of the Wigner distribution is similar to the behavior of the membrane potential in the neuron, in mathematics, this negative point values do not have any meaningful interpretation. We have to understand the oscillation properties of the neuron membrane and this is translated to understand the oscillation properties of the Wigner distribution W_f. A solution for the negative values of the membrane, one might take averages at each point. The standard averaging procedure in mathematics is the convolution of W_f with a smoothing function σ which is centered at the origin, then the convolution can be seen as a local average W_f at (x,w). Where as we already explained, the ω and x signals are the information from the frequency of action potentials and the voltage of the membrane respectively. Since only regions in phase space of area $\delta_x \delta_\omega \geq 1$ are relevant, one may conjecture that the oscillations cancel out and that the convolution is non-negative for all functions whenever the support sigma is large enough.

It is important to find the kernels σ for which the averaged Wigner distribution is always positive. If we take into account this assumption and intuition of the synaptic wave as a Gaussian σ, then the convolution works, and we can consider the average of each point. Now, connected with the circle and the light spectrum just a few millimeters from the edge, with white light in the center, and the coefficient indexes of each color defining different diameters for the cone of light.

Considering the synaptic wave as a general Gaussian has important implications in the sense that we can find all the signal from the general form, and from this kernel recover W_f. Let me refer to the following paper for more details on the step from the uncertainty principle to the Wigner distribution based on the work of A.J.E.M. Janssen [11].

8. Conclusions

Synapses are the target of most, if not all, brain disorders. Because of the enormous number of synapses in our brain (approximately 100,000,000,000), it is often difficult to comprehend how defects in one type of synapse yield the overall loss in brain performance characteristic of all brain disorders. Computational modeling of synapses has been a slow process that is constrained by advances in the interdisciplinary fusion of physics, both for measuring and data analysis, and advances in molecular biology or biophysics. These constraints are also computational: some models have detailed sets of coupled equations with experimental parameters that, although accurate, are impossible to solve because of the enormous number of synapses. This limitation impedes understanding of brain information processing, both in health and in disease states.

Modeling synapses as oscillators with memristors seeks to overcome this computational limitation via an entirely new theoretical approach based on modeling synapses via a wave-particle viewpoint that is computationally efficient. This framework is based on considering synapses as amplitudes of wave activity that vanish or emanate depending on the collective oscillatory behavior. This is a novel way for modeling interneuron communication that does not reside on the complexity of detailed biophysical neuronal models or its synapses. As future consequences of the theoretical proposal, it is necessary to analyze the following implications once experiments at the Amstrong scale are available:

- Analysis of ion channels as wave perturbations and analysis of energy constraints considering oscillations in the Glutamate cycle in interaction with astrocytes for new carriers of long-range correlations, allowing the generation of long-range pattern waves.

- Analysis of mesoscopic oscillations and connection with macroscopic oscillations by applying quantum field theory in the analysis of long-range correlation.

- Improvement of the model considering perturbation details of the arriving wave function.

Perhaps the most valuable contribution of this theoretical proposal is the fact that these equations have an implicit noise generation that allows us to compute under uncertainty conditions without the need for complex stochastic equations for Brownian motion. This can be applied to computation tasks in hardware and software.

Acknowledgements

The author would like to thank the Hoursec team for their motivation to work in a new computational paradigm that will impact the lives of many. The author also thanks the support of our partners and early market traction that are aware of the environmental problem of conventional computing architectures.

Conflict of interest

This chapter should be published under a CC BY License. This publication is copyrighted by the author and subject to all applicable copyright and other protections under U.S. law or other countries regulations. Altering, reselling, redistributing, publishing, or republishing text, figures, data, multimedia, or other information from this copyrighted publication, with the exception of brief, appropriately cited quotations, in any form or medium without prior written permission from the author is prohibited. Systematic downloading, and/or the making of print or electronic copies for public transmission are prohibited. All rights not expressly granted are reserved to the author. The author shall not be required to distribute, and the Subscriber/User shall not redistribute, content from this copyrighted publication to a country where the export thereof is prohibited by U.S. law or other countries regulations. Copyright or Creative Commons License notices may not be removed, obscured, or modified.

Author details

Alexandra Pinto
Hoursec B.V., Delft, Netherlands

*Address all correspondence to: alexandra.pinto@hoursec.tech

References

[1] Lee J et al. Brownian circuits: Designs. International Journal of Unconventional Computing. 2016;**12**:341-362

[2] Tanaka G et al. Recent advances in physical reservoir computing: A review. Neural Networks. 2019;**115**:100-123

[3] Mainzer, K. and Chua, L. Local activity principle (2013).

[4] Dale H. Otto loewi. 1873-1961. Biographical Memoirs of Fellows of the Royal Society. 1962;**8**:67-89

[5] Freeman WJ. Neurodynamics: An Exploration in Mesoscopic Brain Dynamics. London: Springer; 2000

[6] Bullock TH. How do brains work?: papers of a comparative neurophysiologist. Boston: Birkh¨auser; 1993

[7] Chua L. Memristor-the missing circuit element. IEEE Transactions on Circuit Theory. 1971;**18**:507-519

[8] Katchalsky A. Thermodynamics of flow and biological organization. Biophysics and Other Topics. 1976: 521-547

[9] Li Z. A model of olfactory adaptation and sensitivity enhancement in the olfactory bulb. Biological Cybernetics. 1990;**62**:349-361

[10] Nicolis G, Prigogine I. Irreversible processes at nonequilibrium steady states and Lyapounov functions (stability theory/nonequilibrium thermodynamics/fluctuation theory). Physics. 1979;**76**:6060-6061

[11] Janssen A, Leeuwaarden J. Cumulants of the maximum of the gaussian random walk. Stochastic Processes and their Applications. 2007; **117**:1928-1959

Chapter 6

Martingales and Partial Differential Equations Price Valuation Models for European Put Options

Hamilton C. Chinwenyi, Hussaini D. Ibrahim and Theophilus Danjuma

Abstract

In financial mathematics, options are seen as financial transactions that gives the holder the right rather than obligations to buy or sell some specific quantity of an asset in the eminent future at a static price often called the strike price on or before the expiration date of the option contract. In this research work, we examined a typical model in finance, Constant Elasticity of Variance (CEV). Having derived its respective Stochastic Differential Equations (SDEs), we obtained the various Martingales and Partial Differential Equations (PDEs) option price valuation formulas. These were done by using the replicating and riskless portfolio methods. Also, we were able to establish the equivalence of the Martingales and the PDEs methods for European put option pricing for the two different SDE models considered in this research. We, then applied the Girsanov theorem, Martingale theorem as well as Feynman-Kac theorem. Results obtained, show that using our approach described in this research work, we can actually prove the equivalence of the two (Martingales and PDEs) methods by first, beginning with the Martingales option price valuation formula and finally arrive at the Black-Scholes parabolic PDEs in converse.

Keywords: options, financial derivatives, Girsanov theorem, Stochastic Differential Equation, Ito formula, Partial Differential Equations

1. Introduction

In finance, derivatives are usually tools that are attached to a specific asset and through which definite risks in finance can be disposed or handled in financial markets. Various transactions in form of derivatives ought to be treated separately. The value of derivatives in finance is often obtained from the exact prices of the fundamental asset value. Since the future reference price of the underlying item is not known with certainty, the value of the financial derivative at maturity can only be predicted. These derivatives in finance, can be applied for various reasons such as arbitrage, hedging, etc. [1].

Options are one of the examples of derivatives in finance. They are one of the most vital financial instruments. Their recent applications in hedging and arbitrage have earned it widespread attention. There are basically four of its kinds; European option, Asian option, American Option and barrier option [2]. Many reasons may inspire an investor to trade in options rather than trading on the commodities or stocks directly. One of such key reasons is that options save transactions and aids an investor in avoiding risks and market restrictions and fluctuations. Trading options in conjunction with their stock portfolios aids investors in carefully adjusting the risk and return of their investments. Since its introduction in financial mathematics, by Black, Scholes and Merton, the interest in derivatives trading has greatly increased [3].

Determining the values of financial derivatives such as options has since constituted a major problem in the field of financial mathematics. There are two key approaches to derive valuation formulas for a specific financial derivative such as options. These are the Martingales and Partial Differential Equations (PDEs), approaches [4]. In the Martingale approach, the replicating portfolio which is obtained from underlying assets is expected to be self-financing. Since there is no arbitrage, the identified numeraire that converts the replicating portfolio into martingales is used. Hence, value of the derivative is the expectation of the value of the payoff at expiration.

In the case of the PDE approach, a riskless portfolio is formed with the elimination of the stochastic component such that the Stochastic Differential Equation turns to a PDE. A boundary condition is then specified and the resulting PDE is solved by applying the Feynman-Kac theorem. Let us state here that the two approaches as described above are equal [5], even though they both generate different option price valuation formulas; but showing a detailed proof for this is often a difficult task.

This chapter will be centered on the use of advanced mathematical tools (Girsanov and Feynman-Kac Theorems) to carry out a comparative analysis of the Martingale and PDE approaches in the pricing of options using Stochastic Differential Equation (SDE) model in finance as practical examples.

1.1 Definition of terms

In a market economy, we assume there are N assets with values $S_1(t), \cdots, S_N(t)$ at time t, each of which follows a stochastic process.

1.1.1 Option value

Option values are often a function of the underlying asset and time, $u_t = f(S_t, t)$. The calculation of the price of an option (Premium) is our primary concern. For call option, the fundamental value is given by $(S_t - K, 0)$ and for put option, it is given as $(K - S_t, 0)$, $0 \leq t \leq T$. This is actually the value representing the profit an investor gains by exercising his right on the option. The call value is the difference between the underlying price and its fundamental value [6].

1.1.2 Options and replication

Payoff, V_T at time T of an option is always a function of the underlying asset. We look out for a self-financing trading strategy that replicates the same payoff such that $\prod_T = V_T$. This implies a replicating strategy and replicating portfolio [7].

1.1.3 Self-financing portfolio

A portfolio allocation $(\xi_t, \eta_t)_{t\epsilon\mathbb{R}_+}$ with price (value) V_t given by

$$V_t = \xi_t S_t + \eta_t A_t, \quad t\,\epsilon\,\mathbb{R}^+ \tag{1}$$

is self-financing if it satisfies the relation

$$dV_t = \eta_t dA_t + \xi_t dS_t \tag{2}$$

where ξ_t is the stock shares in S_t and η_t is the deposit in the bank.

1.1.4 Risk-neutral measure

A probability measure $\mathbb{P}^*$ on a sample space, Ω is called a risk-neutral measure if it satisfies

$$\mathbb{E}^*\left[S_t|\mathcal{F}_u\right] = e^{r(t-u)}S_u, \quad 0 \le u \le t \tag{3}$$

Where $\mathbb{E}^*$ denotes the expectation under $\mathbb{P}^*$.

1.1.5 Numeraires

Numeraires are assets with positive price, $N_t > 0$, for all given t. The comparative price $\tilde{S}_t$ of an underlying asset is the ratio of the stock price S_t to its numeraire price N_t, i.e. $\tilde{S}_t = \frac{S_t}{N_t}$ and S is usually valued in terms of measured in units of N.

1.1.6 Equivalent martingale measure

Let $\mathbb{Q}$ and $\mathbb{P}$ be probability measures on a sample space Ω. A probability measure, $\mathbb{Q}$ is said to be an an Equivalent Martingale Measure (EMM) for a numeraire N_t, if the following conditions are true:

1. $\mathbb{Q} \sim \mathbb{P}$ and,

2. The relative price $\tilde{S}_t$ is a martingale under $\mathbb{Q}$. This implies

$$E^{\mathbb{Q}}\left[\frac{S_T}{N_T}\middle|\mathcal{F}_t\right] = \frac{S_t}{N_t} \text{ for } T > t. \tag{4}$$

For a complete market economy, given a numeraire N, one can obviously find a unique EMM $\mathbb{N}$ such that asset prices discounted by N are martingale under $\mathbb{N}$. On the contrary, given an EMM, $\mathbb{N}$ we can always find an exclusive numeraire N so that asset prices discounted by N are also, martingale under $\mathbb{N}$.

1.1.7 Martingale

An integrable process $(X_t)_{t\epsilon\mathbb{R}^+}$ is said to be a martingale with respect to the filtration $(\mathcal{F}_t)_{t\epsilon\mathbb{R}^+}$ if

$$E[X_t|\mathcal{F}_s] = X_s, \quad 0 \le s \le t. \tag{5}$$

1.1.8 Fundamental theorem of arbitrage

This Theorem states that if the market is complete, then for each and every numeraire N_t there exists a unique Equivalent Martingale Measure $\mathbb{N}$ such that the relative price of the assets using that numeraire, is a martingale. Therefore, $\frac{S_t}{N_t}$ is a martingale under $\mathbb{N}$. Hence

$$E^{\mathbb{N}}\left[\frac{S_T}{N_T}|\mathcal{F}_t\right] = \frac{S_t}{N_t} \tag{6}$$

If we choose another numeraire M_t, then $\frac{S_t}{M_t}$ is no longer a martingale under $\mathbb{N}$, but there exists another unique equivalent martingale measure, $\mathbb{M}_t$ so that $\frac{S_t}{M_t}$ is a martingale under $\mathbb{M}$. Hence,

$$E^{\mathbb{M}}\left[\frac{S_T}{M_T}|\mathcal{F}_t\right] = \frac{S_t}{M_t} \tag{7}$$

Other financial terms relevant to the discussions in this research work are defined as follows:

Asset (e.g. Stock, interest rates, commodities, etc.): these are objects that provides a claim to future cash flow.

Strike price: this is defined as a fixed price for exercising rights on an option.

Portfolio: a portfolio (or stock portfolio) is an investment in several assets at the same time. The idea of holding portfolio is that it is expected that this reduces the risk in investment.

2. Methods

2.1 Assumptions, parameters and variables of option price models

2.1.1 Assumptions

The assumptions used in the solutions of the model are as follows:

i. All assets (stock) considered this research work are dividends free.

ii. There exists a riskless investment portfolio with a guaranteed return on investment and no chance of default.

iii. There exists no arbitrage in the market which implies a complete market economy.

2.1.2 Parameters and variables

The value of the option depends on the following parameters and variables:
$S(t), X(t), X_t$ or S_t : the underlying asset (stock price) at time t.

σ : the volatility of the underlying asset (measure of the standard deviation of the returns of the asset).

K : the exercise or strike price.

T : time of expiry.

dt : step-size.

r : interest rate.

μ : average rate of growth of the asset.

dS or dX : change in stock price (always positive).

2.2 Preliminaries for model derivation

2.2.1 Girsanov theorem

The basic characteristics of the Girsanov theorem is that of the exponential Martingale

$$Z_t = exp\left(-\frac{1}{2}\int_0^t \theta_s^2 ds + \int_0^t \theta_s dW_s\right). \tag{8}$$

Then, the Girsanov Theorem is stated as follows:

Theorem 2.1 (Girsanov Theorem) [8]: The process $\tilde{W}_t = W_t + \int_0^t \phi_s ds$ is Brownian motion under the measure $\mathbb{Q}$.

Theorem 2.2: Let $(\phi_t)_{t\epsilon[0,T]}$ be a stochastic process satisfying the Novikov integrability condition

$$E\left[exp\left(\frac{1}{2}\int_0^T |\phi|^2 dt\right)\right] < \infty$$

and let $\mathbb{Q}$ denote the probability measure defined by

$$\frac{d\mathbb{Q}}{d\mathbb{P}} = exp\left(-\int_0^T \phi_s dW_s - \frac{1}{2}\int_0^T \phi_s^2 ds\right)$$

then

$$\tilde{W}_t = W_t + \int_0^t \phi_s ds, t \,\epsilon\, [0, T],$$

Is certainly a Brownian motion under the measure, $\mathbb{Q}$.

Theorem 2.3 (Martingale Representation Theorem) [8]: Assuming M_t is an $\mathcal{F}_t-$martingale where $\{\mathcal{F}_t\}_{t\geq 0}$ is the filteration generated by the n-dimensional Brownian motion, $W_t = \left(W_t^{(1)}, \dots, W_t^{(n)}\right)$. If $E\left[M_t^2\right] < \infty$ for all t then there exists a unique n-dimensional adapted stochastic process, ϕ_t such that

$$M_t = M_0 + \int_0^t \phi_s^T dW_t \text{ for all } t \geq 0$$

where ϕ_s^T denotes the transpose of the vector, ϕ_s.

Lemma 2.1 [9]: Let $(\xi_t, \eta_t)_{t\epsilon\mathbb{R}_+}$ be a portfolio strategy with value

$$V_t = \eta_t A_t + \xi_t S_t, t \,\epsilon\, \mathbb{R}^+$$

The following statements are equivalent:
(i) the portfolio strategy $(\xi_t, \eta_t)_{t\epsilon\mathbb{R}^+}$ is self-financing,
(ii)

$$\tilde{V}_t = \tilde{V}_0 + \int_0^t \xi_u dX_u, t \,\epsilon \tag{9}$$

2.2.2 Ito formula

Let us consider the general Ito's formula which is applicable to Ito processes of the type [8]

$$X_t = X_0 + \int_0^t \mu_s ds + \int_0^t \sigma_s dW_s, t \,\epsilon\, \mathbb{R}^+ \tag{10}$$

or in differential notation

$$dX_t = \mu_t dt + \sigma_t dW_t$$

where $(\mu_t)_{t\epsilon\mathbb{R}^+}$ and $(\sigma_t)_{t\epsilon\mathbb{R}^+}$ are square-integrable adapted processes.

Lemma 2.2: For any Ito process $(X_t)_{t\epsilon\mathbb{R}^+}$ of the form Eq. (9) and any $f\epsilon C^{1,2}(\mathbb{R}^+ \times \mathbb{R})$ and $Z_t = f(t, X_t)$ we have,

$$Z_t = f(0, X_0) + \int_0^t \mu_s \frac{\partial f}{\partial x}(s, X_s)ds + \int_0^t \sigma_s \frac{\partial f}{\partial x}(s, X_s)dB_s + \int_0^t \frac{\partial f}{\partial t}(s, X_s)ds + \frac{1}{2}\int_0^t |\sigma_s|^2 \frac{\partial^2 f}{\partial x^2}(s, X_s)ds$$

or in differential form

$$\begin{aligned} dZ_t &= \frac{\partial f}{\partial t}(t, X_t)dt + \frac{\partial f}{\partial x}(t, X_t)dX_t + \frac{1}{2}\frac{\partial^2 f}{\partial x^2}(t, X_t)(dX_t)^2 \\ &= \left(\frac{\partial f}{\partial t}(t, X_t) + \frac{\partial f}{\partial x}(t, X_t)\mu_t + \frac{1}{2}\frac{\partial^2 f}{\partial x^2}(t, X_t)\sigma_t^2\right)dt + \frac{\partial f}{\partial x}(t, X_t)\sigma_t dW_t. \end{aligned} \tag{11}$$

2.2.3 The Black-Scholes PDE

The Black-Scholes Partial Differential Equation (PDE) for the price of a self-financing portfolio is given as:

$$rg(t, x) = \frac{\partial g}{\partial t}(t, x) + rx\frac{\partial g}{\partial x}(t, x) + \frac{1}{2}\sigma^2 x^2 \frac{\partial^2 g}{dx^2}(t, x), x > 0, t \,\epsilon\, [0, T]$$

Proposition 2.1 [9]: Let $(\xi_t, \eta_t)_{t\epsilon\mathbb{R}^+}$ be a portfolio strategy such that.

i. $(\xi_t, \eta_t)_{t\epsilon\mathbb{R}_+}$ is self-financing,

ii. the value $V_t = \eta_t A_t + \xi_t S_t, t \,\epsilon\, \mathbb{R}^+$, takes the form

$$V_t = g(t, S_t)$$

for some $g\epsilon\, C^{1,2}\ ((0,\infty)\times(0,\infty))$.
Then the function $g(t,x)$ satisfies the Black-Scholes PDE

$$rg(t,x) = \frac{\partial g}{\partial t}(t,x) + rx\frac{\partial g}{\partial x}(t,x) + \frac{1}{2}\sigma^2 x^2 \frac{\partial^2 g}{dx^2}(t,x), x > 0, t\,\epsilon\,[0,T] \tag{12}$$

and ξ_t is given by

$$\xi_t = \frac{\partial g}{\partial x}(t, S_t), t\,\epsilon\,\mathbb{R}^+.$$

2.2.4 The Black-Scholes European option price models

In a Black-Scholes European option price models, the concept of arbitrage allows one to establish relationships between prices and hence to determine them. A primary step which determines the theory of valuation of options is the issue of arbitrage. In the real sense of finance, there no prospects of making riskless profit but in most cases in finance, we assume the possibilities of riskless investment which assures return on investment with no choice of uncertainties. Such assumptions can be envisaged in situations of cash deposits in banks or in cases of government bond.

2.2.5 European call and put options

Financial derivatives such as options are financial gears that provides the option holder the right, but not compulsion, for a financial transaction of an underlying asset at a specific time and price [10]. The payoff function of the European Call Option with strike price K, is given by $F(x) = (x-K)^+$ and the Black -Scholes PDE reads

$$\begin{cases} rg_c(t,x) = \dfrac{\partial g_c}{\partial t}(t,x) + rx\dfrac{\partial g_c}{\partial x}(t,x) + \dfrac{1}{2}\sigma^2 x^2 \dfrac{\partial^2 g_c}{\partial x^2}(t,x) \\ g_c(T,x) = (x-K)^+ \end{cases} \tag{13}$$

As earlier stated, call options provide the buyers the right to buy assets at a fixed price, K at the maturation time of the option. At a maturation times, if the asset $(x) > K$; the buyer gains instantaneous profit, $x - K$. However, if $(x) < K$; the buyer refuses to execute the option.

Similarly, for European put options, the Black-Scholes PDE is given as

$$\begin{cases} rg_p(t,x) = \dfrac{\partial g_p}{\partial t}(t,x) + rx\dfrac{\partial g_p}{\partial x}(t,x) + \dfrac{1}{2}\sigma^2 x^2 \dfrac{\partial^2 g_p}{\partial x^2}(t,x) \\ g_p(T,x) = (K-x)^+ \end{cases} \tag{14}$$

For the put option, the option buyer has the right to sell the asset at a agreed price at the maturation date. At option expiry tine, if $(x) < (K)$; then the buyer gains in a profit, $K - x$. But if $(x) > (K)$, he rescinds his decision to execute his right on the option.

2.3 Description of martingale and PDEs options price valuation approach

2.3.1 Martingale approach

In this approach, options are not part of the traded assets $S_1(t), \dots, S_N(t)$, so cannot be priced directly. However, using the arguments in Section 1.1.3, we can form a replicating portfolio $\prod(t) = \sum_{i=1}^{N} a_i(t) S_i(t)$ that reproduces the option price at maturation time, such that $V_t = \Pi_t$ at $t > 0$ as well as $V_T = \Pi_T$. However, theorem of Arbitrage states with a numeraire N_t, each comparative price of the asset is a Martingale relative to the measure $\mathbb{N}$, as well as V_t/N_t as they are blend of Martingale. This implies that V_t/N_t becomes

$$E^{\mathbb{N}}\left[\frac{V_T}{N_T}\middle|\mathcal{F}_t\right] = \frac{V_t}{N_t} \tag{15}$$

from which the time, t price of the option, V_t, is

$$V_t = N_t E^{\mathbb{N}}\left[\frac{V_T}{N_T}\middle|\mathcal{F}_t\right] \tag{16}$$

In the Black-Scholes economy, we have two assets, a stock S_t that follows the Stochastic Differential Equation (SDE)

$$dS = rSdt + \sigma SdW \tag{17}$$

and a fixed bond B such that Eq. (17) is re-written as

$$dS_t = rdt + \sigma S_t dW_t \tag{18}$$

and

$$dB_t = rB_t dt.$$

Applying Girsanov's theorem, the process for dS_t becomes

$$dS_t = rS_t dt + \sigma S_t dW_t^{\mathbb{B}} \tag{19}$$

where $dW_t^{\mathbb{B}} = dW_t + \frac{\mu - r}{\sigma} dt$.

It is straightforward to solve for $B_t = \exp\left(\int_0^t r du\right) = e^{rt}$. We apply Ito's Lemma (Lemma 2.2) to the function $ln\, B_t$. Then $\ln B_t$ follows the SDE

$$d \ln B_t = r_t dt.$$

Integrating from 0 to t we have,

$$\ln B_t - \ln B_0 = \int_0^t r_u du$$

Taking exponential of both sides, so the solution to the SDE is $B_t = \exp\left(\int_0^t r du\right)$ since the time zero value of the bond is $B_0 = 1$. When interest rates are constant then

$r_t = r$ and $B_t = e^{rt}$. Hence, we apply B_t to be the numeraire such that $\tilde{S}_t = \frac{S_t}{B_t}$ becomes a Martingale under the measure, $\mathbb{B}$. The payoff for European Put is given as $V_T = (K - S_T)^+$, hence, by Eq. (16), the Put price at time-t price is

$$V_t = B_t E^{\mathbb{B}}\left[\frac{(K - S_T)^+}{B_T}\bigg|\mathcal{F}_t\right] = e^{-r(T-t)} E^{\mathbb{B}}\left[(K - S_T)^+|\mathcal{F}_t\right] \tag{20}$$

We can either evaluate this expectation directly or obtain the solution for the European put option price.

Alternatively, when we chose a different numeraire. B_t, is subjective, and we can use S_t. We have seen earlier that $\frac{S_t}{B_t}$ is already a martingale under $\mathbb{B}$. Therefore, $\frac{B_t}{S_t}$ would be Martingale, under $\mathbb{S}$. In Eq. (20) the value V_t of the Put is derived from

$$\frac{V_t}{B_t} = E^{\mathbb{B}}\left[\frac{V_T}{B_T}\bigg|\mathcal{F}_t\right]$$

Equivalently, using S_t as the numeraire, the same value V_t can be derived from

$$\frac{V_t}{S_t} = E^{\mathbb{S}}\left[\frac{V_T}{S_T}\bigg|\mathcal{F}_t\right]$$

The European Put has payoff $V_T = (K - S_T)^+$, so the time-t price of the Put is

$$V_t = S_t E^{\mathbb{S}}\left[\left(\frac{K}{S_T} - 1\right)^+\bigg|\mathcal{F}_t\right] = S_t E^{\mathbb{S}}\left[\frac{(K - S_T)^+}{S_T}\bigg|\mathcal{F}_t\right] \tag{21}$$

Even though the expression in Eqs. (20) and (21) are different, they both produce the same solution because the change in numeraire does not affect the option values once an Equivalent Martingales Measure (EMM) has been defined. So, $S_t = B_t = \exp\left(\int_0^t r du\right) = e^{rt}$ which makes equation Eqs. (20) and (21) produce the same solution.

2.3.2 PDE approach

Financial portfolios are pairs of adapted stochastic processes such that at date, t they can only depend on the path of a Brownian motion at time, t [11]. In this approach, Stochastic Differential Equations (SDEs) are converted into a Partial Differential Equation (PDE) with a set boundary condition. The stochastic components of the SDEs are removed and the PDEs becomes easier to solve. A set of initial values or boundary conditions are often provided to have a unique solution. Whereas, the PDE does not give exact solutions, it can then be solved numerically [12].

It is important also to state here that as we shall see later, the Feynman-Kac Theorem (Theorem 2.4 or 2.5), is an essential tool for the pricing of options using the PDE approach [13]. We shall consider an existing model in mathematical finance and then proceed to solve them analytically by applying the two approaches, Martingale and PDEs methods to obtain options price formulas. The SDE model of interest in this chapter is that of Constant Elasticity of Variance (CEV). In this model (CEV), the volatility is assumed to be constant.

2.4 Derivation of options price valuation formula

2.4.1 The Constant Elasticity of Variance (CEV) SDE model

Here, we examine European put option pricing in the CEV SDE model Cox (1975). In this case, we have two assets; the cash deposit in the bank account B is given by $B_t = e^{rt}$ and the stock X follow the SDE

$$dX_t = rX_t dt + \sigma X_t^{\alpha} dW_t, X_0 > 0 \tag{22}$$

with constants $\sigma > 0$ and $\alpha > 0$ where W_t is a standard Brownian motion, $t \in [0, \infty]$, [14]. We proceed to compute the Martingale and PDE options price valuation formula for $u(t, X_t)$.

In order to compute the Martingale option price valuation formula for the SDE model in Eq. (22), we assume $\phi = (\eta_t, \xi_t)_{t\, \epsilon [0,T]}$ be portfolio strategy with price,

$$V_t(\phi) = \eta_t B_t + \xi_t \tag{23}$$

and that it satisfies the self-financing condition

$$dV_t(\phi) = \eta_t dB_t + \xi_t dX_t = re^{rt}\eta_t dt + \xi_t d \tag{24}$$

or equivalently

$$V_t(\phi) = V_0(\phi) + \int_0^t \eta_t dB_s + \int_0^t \xi_t dX_s \tag{25}$$

We proceed by applying notion of change in numeraire. Certainly, the asset prices are strictly positive, that is $B_t > 0$, then the market can be normalized by considering

$$\tilde{B}_t = B_t^{-1} B_t = 1$$

and

$$\tilde{X}_t = B_t^{-1} X_t = e^{-rt} X_t$$

Thus, market normalization implies taking the price B_t of the riskless investment as the numeraire (unit of price) and then calculating other prices in terms of this numeraire.

Hence, the discounted portfolio becomes

$$\tilde{V}_t(\phi) = B_t^{-1} V_t(\phi) = e^{-rt}(\eta_t B_t + \xi_t X_t) = \eta_t + \xi_t \tilde{X}_t$$

and applying integration by parts, we have

$$d\tilde{V}_t(\phi) = B_t^{-1} dV_t(\phi) - re^{-rt} V_t(\phi) dt + \left(dB_t^{-1}\right)(dV_t(\phi)).$$

Since $\left(dB_t^{-1}\right)(dV_t(\phi)) = 0$. Hence, we have that,

$$d\tilde{V}_t(\phi) = B_t^{-1} dV_t(\phi) - re^{-rt} V_t(\phi) dt \tag{26}$$

from Eq. (24), it is known that

$$dV_t(\phi) = \eta_t dB_t + \xi_t dX_t,$$

i.e., ϕ is self-financing, then

$$\begin{aligned}
d\tilde{V}_t(\phi) &= e^{-rt}\{r\eta_t e^{rt}dt + \xi_t dX_t\} - r(\eta_t + \xi_t\tilde{X}_t)dt \\
&= re^{-rt}e^{rt}\eta_t dt + e^{-rt}\xi_t dX_t - r\eta_t dt - r\xi_t e^{-rt}X_t dt \\
&= r\eta_t dt + e^{-rt}\xi_t dX_t - r\eta_t dt - r\xi_t e^{-rt}X_t dt \\
d\tilde{V}_t(\phi) &= \xi_t e^{-rt}dX_t - r\xi_t e^{-rt}X_t dt = \xi_t\{e^{-rt}dX_t + d(e^{-rt})X_t\} \\
d\tilde{V}_t(\phi) &= \xi_t d\tilde{X}_t
\end{aligned}$$

which yields that $\tilde{V}_t(\phi)$ is self-financing because $d\tilde{B}_t = d(1) = 0$. Using Eq. (26) and assuming that $d\tilde{V}_t(\phi) = \xi_t d\tilde{X}_t$ (this implies that ϕ is a self-financing portfolio). Financially speaking, this implies that changes in portfolio values depends only on changes in the prices or values of assets and not on units in which we measure the asset prices.

In integral form, we recall that a discounted market, is written as

$$\tilde{V}_t(\phi) = \tilde{V}_0(\phi) + \int_0^t \xi_s d\tilde{X}_s, t \in \mathbb{R}^+ \tag{27}$$

Next, we need to show that Eq. (27) is a Martingale under $\mathbb{B}$. Then, by Girsanov's Theorem, we can define a probability measure $\mathbb{B}$ and the process

$$\tilde{W}_t = \frac{\mu - r}{\sigma}t + W_t,$$

is a Brownian motion under $\mathbb{B}$.

Now from Eq. (22) if we now compute $d\tilde{X}_t$, we get

$$\begin{aligned}
d\tilde{X}_t &= d(e^{-rt}X_t) = -re^{-rt}X_t dt + e^{-rt}dX_t \\
&= -r\tilde{X}_t dt + e^{-rt}\left[rX_t dt + \sigma X_t^\alpha d\tilde{W}_t\right] \\
&= -r\tilde{X}_t dt + re^{-rt}X_t dt + \sigma X_t^\alpha e^{-rt}d\tilde{W}_t \\
&= -r\tilde{X}_t dt + r\tilde{X}_t dt + \sigma X_t^\alpha e^{-rt}d\tilde{W}_t \\
d\tilde{X}_t &= \sigma\tilde{X}_t^\alpha d\tilde{W}_t
\end{aligned}$$

or unambiguously, let

$$\ln\left(\frac{\tilde{X}_t}{\tilde{X}_0}\right) = \sigma\tilde{X}_t^\alpha d\tilde{W}_t$$

We proceed to look for the solution of the preceding equation using Ito's formula. let,

$$Y(t) = \ln\tilde{X}_t^\alpha, f_t = \frac{\partial\left(\ln\tilde{X}_t^\alpha\right)}{\partial t} = 0, f_x = \frac{\partial\left(\ln\tilde{X}_t^\alpha\right)}{\partial\tilde{X}_t^\alpha} = \frac{1}{\tilde{X}_t^\alpha}, f_{xx} = -\frac{1}{\tilde{X}_t^{2\alpha}}$$

Noting that, $u(t) = \mu X_t, v(t) = \sigma\tilde{X}_t^\alpha$ and so we have

$$d\left(\ln \tilde{X}_t^\alpha\right) = \left[0 + 0 \times \frac{1}{\tilde{X}_t^\alpha} + \frac{1}{2}\left(\sigma \tilde{X}_t^\alpha\right)^2\left(-\frac{1}{\tilde{X}_t^{2\alpha}}\right)\right]dt + \frac{1}{\tilde{X}_t^\alpha}\sigma \tilde{X}_t^\alpha d\tilde{W}_t$$

$$\Rightarrow d\left(\ln \tilde{X}_t^\alpha\right) = \frac{1}{2}\left(\frac{-\sigma^2 \tilde{X}_t^{2\alpha}}{\tilde{X}_t^{2\alpha}}\right)dt + \frac{\sigma \tilde{X}_t^\alpha d\tilde{W}_t}{\tilde{X}_t^\alpha}$$

$$d\left(\ln \tilde{X}_t^\alpha\right) = -\frac{1}{2}\sigma^2 dt + \sigma d\tilde{W}_t.$$

Integrating both sides of the above equation and simplifying, we get

$$\ln \tilde{X}_t^\alpha - \ln \tilde{X}_0^\alpha = \int_0^t -\frac{1}{2}\sigma^2 dt + \int_0^t \sigma d\tilde{W}_t$$

$$\ln \tilde{X}_t^\alpha = \ln \tilde{X}_0^\alpha - \frac{1}{2}\sigma^2 t + \sigma \tilde{W}_t$$

since $W_0 = 0$.

Taking exponential of both sides and simplifying, we get

$$\begin{aligned} e^{\ln \tilde{X}_t^\alpha} &= e^{\left[\ln \tilde{X}_0^\alpha - \frac{\sigma^2}{2}t + \sigma \tilde{W}_t\right]} \\ \tilde{X}_t^\alpha &= e^{\ln \tilde{X}_0^\alpha}.e^{\left[-\frac{\sigma^2}{2}t + \sigma \tilde{W}_t\right]} \\ \therefore \tilde{X}_t^\alpha &= \tilde{X}_0^\alpha exp\left[\sigma \tilde{W}_t - \frac{\sigma^2}{2}t\right] \end{aligned} \tag{28}$$

We now go ahead to show that the above solution in Eq. (28) satisfies the properties of a Martingale under the measure, $\mathbb{B}$. First, we find a measure $\mathbb{B}$ which uses B_t as a numeraire in order to turn it to a Martingale. We now have

$$dX_t = r_t X_t dt + \sigma X_t^\alpha dW_t^{\mathbb{B}}$$

where, $W_t^{\mathbb{B}} = W_t + \frac{\mu - r_t}{\sigma}t$, i.e. this is derived from the application of Girsanov theorem.

By using B_t as the unit price (numeraire), the discounted aaset price $\tilde{X}_t = \frac{X_t}{B_t}$ and $\tilde{X}_t$ are martingales. Hence, by the application of Ito's Lemma to $\tilde{X}_t$, we obtain

$$d\tilde{X}_t = \frac{\partial \tilde{X}}{\partial B}dB_t + \frac{\partial \tilde{X}}{\partial X}dX_t \tag{29}$$

Since all terms involving the second order derivatives are zero. i.e. σ is zero when we compare the discounted stock price $\tilde{X}_t = \frac{X_t}{B_t}$ with Eq. (23). Expanding Eq. (29), we have

$$\begin{aligned} d\tilde{X}_t &= -\frac{X_t}{B_t^2}dB_t + \frac{1}{B_t}dX_t \\ &= -\frac{X_t}{B_t^2}(r_tB_tdt) + \frac{1}{B_t}\left(r_tX_tdt + \sigma X_t^\alpha dW_t^{\mathbb{B}}\right) \\ &= -\frac{X_tr_tdt}{B_t} + \frac{1}{B_t}r_tX_tdt + \frac{\sigma X_t^\alpha dW_t^{\mathbb{B}}}{B_t} \\ &\Rightarrow d\tilde{X}_t = \sigma\tilde{X}_t^\alpha dW_t^{\mathbb{B}} \end{aligned}$$

The solution to the SDE is

$$\tilde{X}_t^\alpha = \tilde{X}_0^\alpha exp\left[-\frac{\sigma^2}{2}t + \sigma W_t^{\mathbb{B}}\right].$$

To show that $\tilde{X}_t^\alpha$ is a Martingale under $\mathbb{B}$, we consider the expectation under $\mathbb{B}$ for $s < t$, hence we have,

$$E^{\mathbb{B}}\left[\tilde{X}_t^\alpha|\mathcal{F}_s\right] = \tilde{X}_0^\alpha \exp\left(-\frac{1}{2}\sigma^2 t\right).E^{\mathbb{B}}\left[\exp(\sigma W_t^{\mathbb{B}})|\mathcal{F}_s\right]$$

$$= \tilde{X}_0^\alpha \exp\left(-\frac{1}{2}\sigma^2 t + \sigma W_s^{\mathbb{B}}\right).E^{\mathbb{B}}\left[\exp(\sigma(W_t^{\mathbb{B}} - W_s^{\mathbb{B}}))|\mathcal{F}_s\right]$$

at time s we have that $W_t^{\mathbb{B}} - W_s^{\mathbb{B}}$ is normally distributed with zero mean and variance t as $N(0, t)$ which is identical in distribution to $W_{t-s}^{\mathbb{B}}$ at time zero. Hence, we can write

$$E^{\mathbb{B}}\left[\tilde{X}_t^\alpha|\mathcal{F}_s\right] = \tilde{X}_0^\alpha \exp\left(-\frac{1}{2}\sigma^2 t + \sigma W_s^{\mathbb{B}}\right).E^{\mathbb{B}}\left[\exp(\sigma(W_{t-s}^{\mathbb{B}})|\mathcal{F}_0\right]$$

Statistically, the moment generating function (mgf) of a random variable X with normal distribution $N(\mu, \sigma^2)$ is stated as

$$E\left[e^{\phi x}\right] = \exp\left(\mu\phi + \frac{1}{2}\phi^2\sigma^2\right).$$

Under $\mathbb{B}$ we have that $W_{t-s}^{\mathbb{B}}$ is $\mathbb{B}$- Brownian motion and distributed as $N(0, t-s)$. Therefore, the mgf of $W_{t-s}^{\mathbb{B}}$ is

$$E^{\mathbb{B}}\left[\tilde{X}_t^\alpha|\mathcal{F}_s\right] = \tilde{X}_0^\alpha \exp\left(-\frac{1}{2}\sigma^2 t + \sigma W_s^{\mathbb{B}}\right).\exp\left(\frac{1}{2}\sigma^2(t-s)\right)$$

where $\sigma = \phi$ and we can then write

$$\begin{aligned} E^{\mathbb{B}}\left[\tilde{X}_t^\alpha|\mathcal{F}_s\right] &= \tilde{X}_0^\alpha \exp\left(-\frac{1}{2}\sigma^2 t + \sigma W_s^{\mathbb{B}}\right) exp\left(\frac{1}{2}\sigma^2(t-s)\right) \\ E^{\mathbb{B}}\left[\tilde{X}_t^\alpha|\mathcal{F}_s\right] &= \tilde{X}_0^\alpha \exp\left(-\frac{1}{2}\sigma^2 s + \sigma W_s^{\mathbb{B}}\right) \\ \therefore E^{\mathbb{B}}\left[\tilde{X}_t^\alpha|\mathcal{F}_s\right] &= \tilde{X}_s^\alpha \end{aligned}$$

We thus have that

$$E^{\mathbb{B}}\left[\tilde{X}_t^{\alpha}|\mathcal{F}_s\right] = \tilde{X}_s^{\alpha}$$

which shows that $\tilde{X}_t^{\alpha}$ is a $\mathbb{B}$ Martingale.
Hence, we have that

$$\tilde{V}_t(\phi) = \tilde{V}_0(\phi) + \int_0^t \xi_s d\tilde{X}_s = \tilde{V}_0(\phi) + \int_0^t \xi_s \sigma \tilde{X}_t^{\alpha} d\tilde{W}_t$$

and $\tilde{V}_t(\phi)$ is a stochastic integral with regard to the Brownian motion under the measure, $\mathbb{B}$. Applying the integrability condition in Theorem 2.2, yields

$$E^{\mathbb{B}}\left[\int_0^T \left|\xi_t \sigma \tilde{X}_t^{\alpha}\right|^2 dt\right] < \infty$$

Hence, we have shown that $\tilde{V}_t(\phi)$ is a martingale under $\mathbb{B}$.
Now since,

$$\tilde{V}_t(\phi) = \tilde{V}_0(\phi) + \int_0^t \xi_s d\tilde{X}_s \; t \in \mathbb{R}^+$$

is a Martingale under $\mathbb{B}$, it follows from the Martingale properties of $\tilde{V}_t(\phi)$ under $\mathbb{B}$ that,

$$\begin{aligned} \tilde{V}_t(\phi) &= E^{\mathbb{B}}\left[\tilde{V}_T|\mathcal{F}_t\right] \\ &= e^{-rT} E^{\mathbb{B}}[V_T|\mathcal{F}_t] \\ &= e^{-rT} E^{\mathbb{B}}[C|\mathcal{F}_t] \end{aligned} \tag{30}$$

where C is a contingent claim, $u(T, X_T)$.
Note that $\phi = (\eta_t, \xi_t)_{t \in [0,T]}$ hedges the claim C, i.e. we have $V_T = C \Longrightarrow V_T = u(T, X_T)$.
Hence Eq. (30) becomes

$$V_t = e^{-rt}\tilde{V}_t = e^{-r(T-t)} E^{\mathbb{B}}[u(T, X_T)|\mathcal{F}_t]$$

Since the process $(X_t)_{t \in \mathbb{R}_+}$ has the Markov property, the value

$$V_t = e^{-r(T-t)} E^{\mathbb{B}}[\phi(X_T)|\mathcal{F}_t] \tag{31}$$

where $\phi(X_T) = u(T, X_T)$ of the portfolio at $t \in [0, T]$ can be written from Eq. (31) as a function $u(t, X_t)$ of t and X_t. Given the payoff function,

$$u(T, X_T) = (K - X_T)^+,$$

Eq. (31) Becomes

$$u(t, X_t) = e^{-r(T-t)} E^{\mathbb{B}}\left[(K - X_T)^+|\mathcal{F}_t\right]$$

$$u(t, X_t) = E^{\mathbb{B}}\left[e^{-r(T-t)}(K - X_T)^+\right]$$

Alternatively, we have

$$\tilde{V}_t(\phi) = \tilde{V}_0(\phi) + \int_0^t \xi_s d\tilde{X}_s, t \in \mathbb{R}^+$$

We the recall the basic theorem of Asset Pricing which stipulates that a market is arbitrage free if there is in existence, at least one Equivalent Martingale Measure (EMM). If we correlate all these facts with Theorem 2.3, then we have all instruments for asset pricing.

Hence, if ϕ is a hedging strategy for C, we have that

$$C = V_0(\phi) + \int_0^T \eta_t dB_t + \int_0^T \xi_t dX_t$$

and the discounted portfolio will be a replicating portfolio for the claim

$$\begin{aligned} \tilde{C} &= e^{-rt}C, i.e. \\ C &= \tilde{V}_T(\phi) = V_0(\phi) + \int_0^T \xi_t d\tilde{X}_t \end{aligned}$$

it follows from the martingale properties of $\tilde{V}(\phi)$ under $\mathbb{B}$ that

$$E^{\mathbb{B}}\left[e^{-rT}C|\mathcal{F}_t\right] = E^{\mathbb{B}}\left[\tilde{V}_T(\phi)|\mathcal{F}_t\right] = \tilde{V}_t(\phi) = e^{-rt}V_t(\phi)$$

which yields

$$V_t(\phi) = E^{\mathbb{B}}\left[e^{-r(T-t)}C|\mathcal{F}_t\right].$$

Now, if

$$C = \phi(X_T) = u(T, X_T),$$

then the arbitrage free price is given by

$$\pi_t(C) = E^{\mathbb{B}}\left[e^{-r(T-t)}\phi(X_T)|\mathcal{F}_t\right]$$

X_t solves the following SDE

$$dX_t = rX_t dt + \sigma X_t^\alpha d\tilde{W}_t$$

under $\mathbb{B}$. Hence, X_t is a Markov process, and we have that

$$E^{\mathbb{B}}\left[e^{-r(T-t)}\phi(X_T)|\mathcal{F}_t\right] = E^{\mathbb{B}}\left[e^{-r(T-t)}\phi(X_T)\right]_{|x=X_t} \tag{32}$$

given the payoff function,

$$\phi(X_T) = u(T, X_T) = (K - X_T)^+.$$

Hence, Eq. (32) becomes

$$u(t, X_t) = E^{\mathbb{B}}\left[e^{-r(T-t)}(K - X_T)^+\right]$$

Next, we shall derive the PDE valuation formula for the above SDE model in Eq. (22). There are two assets in the model; the bank account B is given by $B_t = e^{rt}$ and the stock X follows the SDE

$$dX_t = rX_t dt + \sigma X_t^\alpha dW_t, X_0 > 0$$

with constants $\sigma > 0$ and $\alpha > 0$, where W is a $Q -$ Brownian motion. Applying Ito's lemma (Lemma 2.2), the value $u = u(X_t, t)$ of the Option written on X follows the SDE

$$du = \left(\frac{\partial u}{\partial t} + rx\frac{\partial u}{\partial x} + \frac{1}{2}\sigma^2 x^{2\alpha}\frac{\partial^2 u}{\partial x^2}\right)dt + \frac{\partial u}{\partial x}\sigma x^\alpha dW \tag{33}$$

We transform the SDE for u into PDE, and define some boundary conditions to allow a solution to the PDE. We then proceed to set up a risk-free self-financing and replicating portfolio $\prod$ consisting of η units of the option and ξ units of the asset, to have

$$d\prod = \eta du + \xi dX$$

Hence, the portfolio trails the SDE

$$\begin{aligned} d\prod &= \eta\left[\left(\frac{\partial u}{\partial t} + rx\frac{\partial u}{\partial x} + \frac{1}{2}\sigma^2 x^{2\alpha}\frac{\partial^2 u}{\partial x^2}\right)dt + \frac{\partial u}{\partial x}\sigma x^\alpha dW\right] + \xi[rxdt + \sigma x^\alpha dW] \\ &= \eta\left(\frac{\partial u}{\partial t} + rx\frac{\partial u}{\partial x} + \frac{1}{2}\sigma^2 x^{2\alpha}\frac{\partial^2 u}{\partial x^2}\right)dt + \eta\sigma x^\alpha\frac{\partial u}{\partial x}dW + \xi rxdt + \xi\sigma x^\alpha dW \\ d\prod &= \left[\eta\frac{\partial u}{\partial t} + \eta rx\frac{\partial u}{\partial x} + \frac{1}{2}\eta\sigma^2 x^{2\alpha}\frac{\partial^2 u}{\partial x^2} + \xi rx\right]dt + \left[\xi\sigma x^\alpha + \eta\sigma x^\alpha\frac{\partial u}{\partial x}\right]dW \end{aligned} \tag{34}$$

We take the choices of $\eta = 1$ and $\xi = -\frac{\partial u}{\partial x}$ which eliminates the stochastic element from $d\prod$ and makes the portfolio risk-free. The portfolio being risk-free, it earns a riskless rate and the return on portfolio investment in small time increment dt becomes

$$d\prod = r\prod dt$$

Substituting for ξ in Eq. (34) and equating it with $r\prod dt$, produces a Black-Scholes PDE for $u(t, X_t)$.

$$\begin{aligned} d\prod &= \left[\eta\frac{\partial u}{\partial t} + \eta rx\frac{\partial u}{\partial x} + \frac{1}{2}\eta\sigma^2 x^{2\alpha}\frac{\partial^2 u}{\partial x^2} - rx\frac{\partial u}{\partial t}\right]dt + \left[-\sigma x^\alpha\frac{\partial u}{\partial x} + \eta\sigma x^\alpha\frac{\partial u}{\partial x}\right]dW = \left(r\eta u - rx\frac{\partial u}{\partial x}\right)dt \\ &\Longrightarrow \eta\frac{\partial u}{\partial t} + \eta rx\frac{\partial u}{\partial x} + \frac{1}{2}\eta\sigma^2 x^{2\alpha}\frac{\partial^2 u}{\partial x^2} - rx\frac{\partial u}{\partial x} = r\eta u - rx\frac{\partial u}{\partial x} \\ &\Longrightarrow \eta\frac{\partial u}{\partial t} + \eta rx\frac{\partial u}{\partial x} + \frac{1}{2}\eta\sigma^2 x^{2\alpha}\frac{\partial^2 u}{\partial x^2} - r\eta u = 0 \end{aligned}$$

but $\eta = 1$, therefore

$$\Longrightarrow \frac{\partial u}{\partial t} + rx\frac{\partial u}{\partial x} + \frac{1}{2}\sigma^2 x^{2\alpha}\frac{\partial^2 u}{\partial x^2} - ru = 0. \tag{35}$$

We have been able to establish a risk-free portfolio with X and u resulting to a PDE since X and u are each determined by one source of ambiguity which is the Brownian motion W. This enhances the possibility of eliminating the stochastic element thereby resulting to a PDE.

However, we then progress to seek a solution to the PDE above by the application of the Feynman-Kac Theorem (Leon [15], Sarkka [16], Wiktorsson [17]) or numerically given a certain boundary condition. We note that the value $u_t = u(X_t, t)$ of European put option at time t having a strike price K with a constant interest rate, r which follows the Black-Scholes PDE as stated in Eq. (35) with boundary condition $(X_T, T) = (K - X_T)^+$.

2.5 Equivalence of the martingales and PDEs valuation formulas for option pricing models

The two stipulated methods for the derivation of option price valuation formulas as earlier discussed are equal. This can be ascertained through a rigorous proof using the examples earlier discussed in this work. The method employed by Heath and Schweizer [4] to show the equivalence of the two approaches is:

i. To prove that the valuation formula obtained through the Martingale method can be solved using Ito's formula, which implies that under some stated verifiable necessary and sufficient conditions that it satisfies the corresponding valuation PDE formula, at least on the support of the underlying diffusion.

ii. To establish that a solution exists for the derived PDE valuation formula.

However, in our research work, we looked at a possibility of establishing the equivalence of the two methods as earlier discussed. This is achieved by the application of Girsanov theorem and Feynman-Kac theorem to drive home our argument that the two approaches are equivalent. That is to show that the Martingale (No-arbitrage) method yields the Black-Scholes PDEs and vice versa.

In financial mathematics, the Feynman-Kac theorem provides the required relationship between parabolic Partial Differential Equations (PDEs) and stochastic processes. It represents solutions of a parabolic PDE forms of conditional expectations. Conversely, an important class of expectations of random processes can be computed by deterministic methods. The theorem can be in one dimension, and in multiple dimensions [12, 16].

Theorem 2.4: The Feynman-Kac Theorem in one dimension

Suppose that x_t follows the stochastic process

$$dx_t = \mu(x_t, t)dt + \sigma(x_t, t)dW_t^{\mathbb{Q}} \tag{36}$$

where $W_t^{\mathbb{Q}}$ is a Weiner process under the measure $\mathbb{Q}$. If, $V(x_t, t)$ is differentiable with respect to x_t and t and $V(x_t, t)$ follows the Partial Differential Equation (PDE) stated as

$$\frac{\partial V}{\partial t} + \mu(x_t, t)\frac{\partial V}{\partial x} + \frac{1}{2}\sigma(x_t, t)^2\frac{\partial^2 V}{\partial x^2} - r(x_t, t)V(x_t, t) = 0 \tag{37}$$

and with boundary condition $V(X_T, T)$. The theorem asserts that $V(x_t, t)$ has the solution

$$V(x_t, t) = E^{\mathbb{Q}}\left[e^{-\int_t^T r(x_u, u)du} V(X_T, T) \middle| \mathcal{F}_t \right] \tag{38}$$

Note that the expectation is taken under the measure $\mathbb{Q}$ that makes the stochastic term in Eq. (36) Brownian motion. The generator of the process in Eq. (36) is defined as the operator

$$A = \mu(x_t, t)\frac{\partial}{\partial x} + \frac{1}{2}\sigma(x_t, t)^2\frac{\partial}{\partial x^2} \tag{39}$$

So, the PDE in $V(x_t, t)$ is sometimes written

$$\frac{\partial V}{\partial t} + AV(x_t, t) - r(x_t, t)V(x_t, t) = 0 \tag{40}$$

The Theorem can be used in two ways:

i. If x_t adapts to the process in Eq. (36) and given a function $V(x_t, t)$ with initial boundary condition $V(X_T, T)$, then we get a solution for $V(x_t, t)$ as in Eq. (38).

ii. If we know that the solution to $V(x_t, t)$ is given by Eq. (38) and that x_t follows the process in Eq. (36), then we are assured that $V(x_t, t)$ satisfies the PDE in Eq. (40).

Theorem 2.5: Multidimensional version of the Feynman-Kac theorem
Suppose that x_t follows the stochastic process in n dimensions

$$dx_t = \mu(x_t, t)dt + \sigma(x_t, t)dW_t^{\mathbb{Q}}$$

where x_t and $\mu(x_t, t)$ are each vector of dimension n, $W_t^{\mathbb{Q}}$ is a vector of dimension m of $\mathbb{Q}$ – Brownian motion, and $\sigma(x_t, t)$ is a matrix of size $n \times m$. In other words

$$d\begin{pmatrix} x_1(t) \\ \vdots \\ x_n(t) \end{pmatrix} = \begin{pmatrix} \mu_1(x_t, t) \\ \vdots \\ \mu_n(x_t, t) \end{pmatrix} dt + \begin{pmatrix} \sigma_{11}(x_t, t) & \cdots & \sigma_{1m}(x_t, t) \\ \vdots & \ddots & \vdots \\ \sigma_{n1}(x_t, t) & \cdots & \sigma_{nm}(x_t, t) \end{pmatrix} \begin{pmatrix} dW_1^{\mathbb{Q}}(t) \\ \vdots \\ dW_m^{\mathbb{Q}}(t) \end{pmatrix}$$

The generator of the process is

$$A = \sum_{i=1}^{n} \mu_i \frac{\partial}{\partial x_i} + \frac{1}{2}\sum_{i=1}^{n}\sum_{j=1}^{n} \left(\sigma\sigma^T\right)_{ij} \frac{\partial^2}{\partial x_i \partial x_j} \tag{41}$$

where for notational convenience $\mu_i = \mu_i(x_t, t), \sigma = \sigma(x_t, t)$, and $\left(\sigma\sigma^T\right)_{ij}$ is components (i, j) in the matrix $\sigma\sigma^T$ of dimensions $(n \times n)$. Hence, the Partial Differential Equation (PDE) in $V(x_t, t)$ is given as

$$\frac{\partial V}{\partial t} + AV(x_t, t) - r(x_t, t)V(x_t, t) = 0 \tag{42}$$

and with boundary condition $V(X_T, T)|\mathcal{F}_t$ has solution

$$V(x_t, t) = E^{\mathbb{Q}}\left[e^{-\int_t^T r(x_u, u)du} V(X_T, T)\middle|\mathcal{F}_t\right] \tag{43}$$

2.5.1 Equivalence of the martingale and PDEs option valuation formulas for constant elasticity of variance (CEV) SDE model

We now proceed to prove the equality of the Martingales and PDEs approaches for the given SDE model in finance.

We consider the CEV SDE model given in Eq. (22) as

$$dX_t = rX_t dt + \sigma X_t^{\alpha} dW_t$$

and the fixed bond is given as

$$dB_t = rB_t dt$$

We apply the Girsanov's Theorem so that the process for dX_t for CEV model in Eq. (22) becomes

$$dX_t = rX_t du + \sigma X_t^{\alpha} dW_t^{\mathbb{B}} \tag{44}$$

The European put has payoff

$$u_T = (K - X_T)^+$$

So in accordance with Eq. (21), the time, t price of the put option is

$$u_t = u(t, X_t) = B_t E^{\mathbb{B}}\left[\frac{K - X_T}{B_T}|\mathcal{F}_+\right]$$

were

$$\begin{aligned} B_t &= exp\left(\int_0^t rdu\right) = e^{rt} \\ u_t &= e^{-r(T-t)} E^{\mathbb{B}}\left[(K - X_T)^+|\mathcal{F}_t\right] \end{aligned} \tag{45}$$

We ascertain that the discounted option price $\frac{u_t}{B_t}$ becomes a martingale only when the Black-Scholes parabolic PDE is fulfilled or solved. In a Black-Scholes model, the arbitrage-free price of an option u_t is given as

$$u_t = N_t E^{\mathbb{N}}\left[\frac{u_T}{N_T}|\mathcal{F}_t\right] \tag{46}$$

with numeraire $N_t = B_t$

$$u_t = u(X_t, t) = e^{-r(T-t)} E^{\mathbb{B}}[u[X_T, T]|\mathcal{F}_t] \tag{47}$$

where $\mathbb{B}$ is the measure under which the discounted stock price

$$\frac{X_t}{B_t} = e^{-rt} X_t$$

is a martingale. Under $\mathbb{B}$, the discounted option price,

$$Z(X_t, t) = e^{-rt} u(X_t, t),$$

is also a Martingale, which we write as

$$Z_t = e^{-rt} u_t$$

for convenience.
By Ito's lemma, $Z(X_t, t)$ follows the process

$$dZ_t = \frac{\partial Z}{\partial t} dt + \frac{\partial Z}{\partial X} dX_t + \frac{1}{2}\frac{\partial^2 Z}{\partial X^2}(dX_t)^2 \tag{48}$$

We need the Ito multiplication table given below
Now by differentiating $Z_t = e^{-rt} u_t$, we get

$$\frac{\partial Z}{\partial t} = -re^{-rt} u_t + e^{-rt}\frac{\partial u}{\partial t}. \tag{49}$$

Substituting Eq. (49) into Eq. (48) and substituting for dX_t for the CEV model in Eq. (22) yields

$$dZ_t = e^{-rt}\left[-ru_t + \frac{\partial u}{\partial t}\right] dt + \frac{\partial Z}{\partial X}\left[rX_t dt + \sigma X_t^{\alpha} dW_t\right] + \frac{1}{2}\frac{\partial^2 Z}{\partial X^2}\sigma^2 X_t^{2\alpha} dt \tag{50}$$

Note that

$$\frac{\partial Z}{\partial X} = e^{-rt}\frac{\partial u}{\partial X} \text{ and } \frac{\partial^2 Z}{\partial X^2} = e^{-rt}\frac{\partial^2 u}{\partial X^2}.$$

Substituting for this in Eq. (50), we have

$$\begin{aligned}
dZ_t &= e^{-rt}\left[-ru_t + \frac{\partial u}{\partial t}\right] dt + rX_t \frac{\partial Z}{\partial X} dt + \frac{\partial Z}{\partial X}\sigma X_t^{\alpha} dW_t + \frac{1}{2}\frac{\partial^2 Z}{\partial X^2}\sigma^2 X_t^{2\alpha} dt \\
dZ_t &= e^{-rt}\left[-ru_t + \frac{\partial u}{\partial t}\right] dt + e^{-rt} rX_t \frac{\partial u}{\partial X} dt + e^{-rt}\sigma X_t^{\alpha}\frac{\partial u}{\partial X} dW_t + \frac{1}{2} e^{-rt}\sigma^2 X_t^{2\alpha}\frac{\partial^2 u}{\partial X^2} dt \\
dZ_t &= e^{-rt}\left[-ru_t + \frac{\partial u}{\partial t} + rX_t\frac{\partial u}{\partial X} + \frac{1}{2}\sigma^2 X_t^{2\alpha}\frac{\partial^2 u}{\partial X^2}\right] dt + e^{-rt}\left[\sigma X_t^{\alpha}\frac{\partial u}{\partial X}\right] dW_t
\end{aligned} \tag{51}$$

The conditional expectation under which the martingale option price is derived is taken under the measure, $\mathbb{B}$. Hence, the process dZ_t must indicate this. Therefore, making a change of the measure to $\mathbb{B}$ in Eq. (51) to obtain

$$dZ_t = e^{-rt}\left[-ru_t + \frac{\partial u}{\partial t} + rX_t\frac{\partial u}{\partial X} + \frac{1}{2}\sigma^2 X_t^{2\alpha}\frac{\partial^2 u}{\partial X^2}\right]dt + e^{-rt}\left[\sigma X_t^{\alpha}\frac{\partial u}{\partial X}\right]\left[dW_t^{\mathbb{B}} - \left(\frac{\mu - r}{\sigma}\right)dt\right] \tag{52}$$

Since $dW_t^{\mathbb{B}} = dW_t + \left(\frac{\mu - r}{\sigma}\right)dt$ from the application of Girsanov's Theorem that produced Eq. (44). Rearranging the terms in Eq. (52), we have

$$dZ_t = e^{-rt}\left[-ru_t + \frac{\partial u}{\partial t} + rX_t\frac{\partial u}{\partial X} + \frac{1}{2}\sigma^2 X_t^{2\alpha}\frac{\partial^2 u}{\partial X^2}\right]dt + e^{-rt}\left[\sigma X_t^{\alpha}\frac{\partial u}{\partial X}\right]dW_t^{\mathbb{B}}. \tag{53}$$

Hence, we have that in Eq. (51)

$$Z_t = e^{-rt}u[X_t, t],$$

the discounted time, t option price, is a martingale under $\mathbb{B}$ only when the drift term in Eq. (53) is zero. That is when

$$\frac{\partial u}{\partial t} + rx\frac{\partial u}{\partial x} + \frac{1}{2}\sigma^2 x^{2\alpha}\frac{\partial^2 u}{\partial x^2} - ru_t = 0.$$

We recognize this equation as the Black-Scholes PDE for the given SDE in Eq. (22). Obviously, it can be said that the price of the option can only be a Martingale if the Black-Scholes parabolic PDE is fulfilled. It suffices to state that the value of u_t that solves the PDE given the boundary condition $u(T, X_T) = (K - X_T)^+$ is the same with the u_t in Eq. (45) and we can see that it is the same equation with the one derived in Eq. (35).

Moreso, we proceed to prove the equivalence using the Feynman-Kac theorem. Now given the derived PDE for the CEV model in Eq. (35), we proceed to solve the PDE by applying the Feynman-Kac theorem to ascertain if we can derive the Martingales formula.

$$\frac{\partial u}{\partial t} + rx\frac{\partial u}{\partial x} + \frac{1}{2}\sigma^2 x^{2\alpha}\frac{\partial^2 u}{\partial x^2} - ru = 0$$

with the boundary condition

$$u(T, X_T) = (K - X_T)^+ = \psi(x).$$

Let $X(t)$ solve the Stochastic Differential Equation

$$dX = rXdt + \sigma X^{\alpha}dW$$

with initial condition $x(t) = x$.

We define a process $x(t)$ on interval $[t, T]$ as follows:

$$dX = rxdt + \sigma x^{\alpha}dW, x(t) = x$$

Applying Ito's formula (Lemma 2.2) on $u(t, X_t)$, we have

$$du = \frac{\partial u}{\partial t}dt + \frac{\partial u}{\partial x}dX + \frac{1}{2}\frac{\partial^2 u}{\partial x^2}(dX)^2$$
$$(dX)^2 = (rxdt + \sigma x^\alpha dW)(rxdt + \sigma x^\alpha dW)$$
$$(dX)^2 = (rx)^2(dt)^2 + \sigma^2 x^{2\alpha}(dW)^2 + 2rx\sigma x^\alpha(dt.dW)$$

Using the Ito multiplication table, **Table 1**, we have

$$(dX)^2 = \sigma^2 x^{2\alpha} dt$$

Therefore

$$\begin{aligned} du &= \frac{\partial u}{\partial t}dt + \frac{\partial u}{\partial x}(rxdt + \sigma x^\alpha dW) + \frac{1}{2}\frac{\partial^2 u}{\partial x^2}\sigma^2 x^{2\alpha}dt \\ &= \frac{\partial u}{\partial t}dt + rx\frac{\partial u}{\partial x}dt + \sigma x^\alpha \frac{\partial u}{\partial x}dW + \frac{1}{2}\frac{\partial^2 u}{\partial x^2}\sigma^2 x^{2\alpha}dt \\ &= \frac{\partial u}{\partial t}dt + rx\frac{\partial u}{\partial x}dt + \frac{1}{2}\frac{\partial^2 u}{\partial x^2}\sigma^2 x^{2\alpha}dt + \frac{\partial u}{\partial x}\sigma x^\alpha dW \\ du &= \left[\frac{\partial u}{\partial t} + rx\frac{\partial u}{\partial x} + \frac{1}{2}\frac{\partial^2 u}{\partial x^2}\sigma^2 x^{2\alpha}\right]dt + \frac{\partial u}{\partial x}\sigma x^\alpha dW \end{aligned} \tag{54}$$

But, $\left[\frac{\partial u}{\partial t} + rx\frac{\partial u}{\partial x} + \frac{1}{2}\frac{\partial^2 u}{\partial x^2}\sigma^2 x^{2\alpha}\right] = ru$ as given in Eq. (35).
So, we now have

$$du = rudt + \frac{\partial u}{\partial x}\sigma x^\alpha dW$$

Integrating from t to T, we have

$$\int_t^T du = \int_t^T rudt + \int_t^T \frac{\partial u}{\partial x}\sigma x^\alpha dW$$

which gives

$$u(x(T), T) - u(x(t), t) = \int_t^T rudt + \int_t^T \frac{\partial u}{\partial x}\sigma x^\alpha dW$$

Substituting the initial and terminal terms, we get

$$\psi(x(T)) - u(x, t) = \int_t^T rudt + \int_t^T \frac{\partial u}{\partial x}\sigma x^\alpha dW$$

•	dt	dW_t
dt	0	0
dW_t	0	dt

Table 1.
Ito's multiplication table.

where $\psi(x(T)) = (K - X_T)^+$

$$\psi(x(T)) - u(x,t) = \int_t^T rudt + \int_t^T \frac{\partial u}{\partial x}\sigma x^\alpha dW. \tag{55}$$

Taking expectations of both sides, we have

$$\begin{aligned} E[\psi(x(T)) - u(x,t)] &= E\left[\int_t^T rudt\right] + E\left[\int_t^T \frac{\partial u}{\partial x}\sigma x^\alpha dW\right] \\ \text{but,} \qquad E\left[\int_t^T \frac{du}{dx}\sigma x^\alpha dB\right] &= 0 \end{aligned} \tag{56}$$

We now have

$$E[u(x,t)] = E\left[\psi(x(T)] - E\left[\int_t^T rudt\right]\right.$$

$$E[u(x,t)] = E\left[\psi(x(T)) - \int_t^T rudt\right] = E^{\mathbb{B}}\left[e^{-\int_t^T rsds}\psi(x(T))|F_t\right]$$

$$u(x,t) = E^{\mathbb{B}}\left[e^{-r(T-t)}(K - X_T)^+|F_t\right]$$

$$\therefore u(x,t) = e^{-r(T-t)}E^{\mathbb{B}}\left[(K - X_T)^+\right]$$

where $u(x,t) = u(X_t, t)$, which is the same option price valuation formula for the CEV model derived in Eq. (47).

3. Conclusion

In this chapter, we have examined the Martingale and PDEs approaches for the valuation of options price for two Stochastic Differential Equation models in finance using the replicating and riskless portfolio methods. The two SDE models studied are the Constant Elasticity of Variance (CEV) model and the Heston stochastic volatility model respectively. The model parameters are given in Section 2.1.2.

From this chapter, the following conclusions were drawn:

i. The Martingales and PDEs approach for the valuation of options price removes the stochastic components in Stochastic Differential Equation (SDE) models, thereby making the models riskless and hence easier for the resulting equations to be solved analytically or numerically.

ii. That adequate understanding of the two approaches in valuation of financial Stochastic Differential Equation (SDE) models would aid an investor in the mitigation of the risk involved in financial derivative pricing and also in determining the quality of the choice (right) the holder of the option makes at the maturation date of the option.

iii. Analytical results obtained through the application of Girsanov's theorem for defining an Equivalent Martingale Measure and Feynman-Kac theorem respectively has shown that the two option price valuation formulas derived by both the Martingales and PDEs approaches are equivalent.

Author details

Hamilton C. Chinwenyi[1*], Hussaini D. Ibrahim[1] and Theophilus Danjuma[2]

1 Raw Materials Research and Development Council, Maitama, Abuja, Nigeria

2 Department of Mathematical Science, Federal University Gusau, Zamfara State, Nigeria

*Address all correspondence to: chinwenyi@yahoo.com

References

[1] IMF. Eleventh Meeting of the International Monetary Fund (IMF) Committee on Balance of Payments Statistics on Financial Derivatives. Washington, D.C.: International Monetary Fund Publications Services; 1998. pp. 14-15

[2] Yang Y. Valuing a european option with the heston model [M.Sc. thesis]. Rochester: School of the Mathematical Sciences, Rochester Institute of Technology; 2013

[3] Gong H, Thavaneswaran A, Liang Y. Recent developments in option pricing. Journal of Mathematical Finance. 2011;**1**: 63-71

[4] Heath D, Schweizer M. Martingales versus PDEs in finance: An equivalence result with examples. A Journal of Applied Probability. 2000;**37**:947-957

[5] Chinwenyi HC, Onah ES, Aboiyar T. The equivalence of martingales and Partial differential equations (PDEs) approach for European put option pricing. Journal of the Nigerian Association of Mathematical Physics. 2017;**42**:61-70

[6] Ntwiga BD. Numerical methods for the valuation of financial derivatives [M. Sc. thesis]. South Africa: Department of Mathematics and Applied Mathematics, University of Western Cape; 2005

[7] Chinwenyi HC, Onah ES, Aboiyar T. Partial differential equation approach for valuation of European put option price using two stochastic financial models. International Journal of Mathematical Analysis and Optimization: Theory and Applications. 2018;**4**(1):276-290

[8] Haugh M. Introduction to stochastic calculus. Journal of Financial Engineering: Continuous-Time Models, Columbia University. 2010;**1**:5-11

[9] Privault N. Notes on Stochastic Finance. Lecture Note 7 on Martingale Approach to Pricing and Hedging. Vol. 2016. Singapore Nanyang Technological University; 2016. pp. 1-42. Available from: https://www.ntu.edu.sg/home/nprivault/MA5182/

[10] Adaji MO, Onah SE, Kimbir RA, Aboiyar T. Financial valuation of early retirement benefit in a defined benefit pension plan. International Journal of Mathematical Analysis and Optimization: Theory and Applications. 2016;**2**(1):67-82

[11] Jeanblanc M. Lectures on Mathematical Finance. Hong Kong: City University; 2001. pp. 25-50

[12] Rouah FD. Mathematical Finance Notes on Four Derivations of the Black Scholes PDE. Vol. 20172017. pp. 1-9. Available from: http://www.frouah.com/finance%20notes/Black%20Schoes%PDE.pdf

[13] Wilcox D. The PDE Methods for Pricing Derivative Securities. University of Cape Town: South African Press; 2005. pp. 67-94

[14] Ogbogbo CP. Modeling crude oil spot price as an Ornstein – Uhlenbeck process. International Journal of Mathematical Analysis and Optimization: Theory and Applications. 2018;**4**(1):261-275

[15] Leon WCH. MFE6516 Stochastic Calculus for Finance. Nanyang Business School, Singapore Nanyang Technological University; 2014. pp. 1-32

[16] Sarkka S. Lecture 5: Further Topics; Series Expansions, Feynman-Kac, Girsanov Theorem, Filtering Theory. Finland: Aalto University, Tampere University of Technology and Lappeenranta University of Technology; 2012. pp. 1-20. Matingale-pricing-hedging.pdf

[17] Wiktorsson M. Lecture Note on some Problems with Solutions. Fmsn25/ Masm24 Valuation of Derivative Assets, Faculty of Engineering. Lund University: Centre for Mathematical Sciences and Mathematical Statistics; 2016. pp. 1-7

Chapter 7

Optimal Stochastic H^∞ Controller Synthesis via LQR Theory

Yung Kuan Foo and Yeng Chai Soh

Abstract

Existing published works on H^∞ and mixed H^∞/H^2 control generally solve the problem via game theory and saddle-point solutions. We provide an alternative solution via the stochastic LQR approach. Specifically, we consider the problem of H^∞ state regulation problem for continuous-time systems which are subjected to both energy-bounded deterministic disturbances and stochastic white noises, and uncertain initial system state $\mathbf{x}(0)$ satisfying $E\{\mathbf{x}(0)\} = \mathbf{0}$ and $E\{\mathbf{x}(0)\mathbf{x}(0)^T\} \leq \mathbf{X}_0$. We derive feedback laws which minimize $\sup_{\|\mathbf{w}(t)\|2 \leq 1} E\{\|\mathbf{\Pi}^{1/2}\mathbf{x}(t)\|_T^{\,2} + \|\mathbf{R}^{1/2}\mathbf{u}(t)\|_T^{\,2}\}$, where $\mathbf{\Pi}$ and $\mathbf{R}$ are positive-definite weight matrices. We further demonstrate how the approach may be extended to solving the robust stochastic LQR control with uncertainty in the system $\mathbf{A}$ matrix.

Keywords: H^∞ control, LQ regulator, stochastic regulator, robust control, mixed H^2/H^∞ control

1. Introduction

Existing approaches to solving H^∞ control problem typically pose the problem as a differential game where the solution is obtained by finding the minimax saddle point; see [1–7] for example. This approach is intuitive from the H^∞ perspective since the problem is indeed mathematically a minimax problem. However, when it is found that the Riccati equations obtained are indeed very similar to the standard Riccati equations obtained from related LQR problems with the same system matrices, tremendous interests on how to reconcile and combine the two approaches in a fruitful way have been generated. This has given rise to the once-very-hotly investigated mixed H^2/H^∞ control problem [1, 8–10]. Specifically, the system model considered is of the form:

$$\begin{aligned} d\mathbf{x}(t)/dt &= \mathbf{A}\mathbf{x}(t) + \mathbf{B_u}\mathbf{u}(t) + \mathbf{E}_2\mathbf{w}_2(t) + \mathbf{E}_\infty\mathbf{w}_\omega(t) \\ \mathbf{z}_2(t) &= \mathbf{C}_2\mathbf{x}(t) + \mathbf{D}_2\mathbf{w}_2(t) \\ \mathbf{z}_\infty(t) &= \mathbf{C}_\infty\mathbf{x}(t) + \mathbf{D}_\infty\mathbf{w}_\infty(t), \mathbf{x}(0) = \mathbf{0} \end{aligned} \tag{1}$$

Let $\mathbf{T}_2$ and $\mathbf{T}_\infty$ denote the closed-loop transfer functions from $\mathbf{w}_2$ to $\mathbf{z}_2$ and from $\mathbf{w}_\infty$ to $\mathbf{z}_\infty$, respectively. The problem was to minimize the H^2 norm of $\mathbf{T}_2$ subject to the H^∞ norm of $\mathbf{T}_\infty$ smaller than certain prescribed positive value γ. Clearly, if $\mathbf{w}_\infty = \mathbf{0}$,

IntechOpen

then the H^2 norm will give a performance measure on the effect of $\mathbf{w}_2$. Conversely, if $\mathbf{w}_2 = \mathbf{0}$, then the H^∞ norm would give a performance bound on the effect of $\mathbf{w}_\infty$. But what if $\mathbf{w}_2$ and $\mathbf{w}_\infty(t)$ are both nonzero? The main focus of such an approach is usually to optimize the H^2 norm for performance, and the H^∞ norm bound was more for "robustness" against model uncertainties than for worst-case performance. Thus, we start with a signal-based worst-case control problem, but eventually end up with a "robust" (bound on the H^∞ norm) LQR control solution. Such an approach is justifiable but an implicit implication is that the performance is only optimized for the H^2 norm and hence *not*, at least not expressly, about worst-case performance optimization. Another fact to note is that though we obtain a solution to guarantee the H^∞ bound of the closed-loop system by solving the relevant Riccati equation (Eq. (36)), such bound is generally not tight. This fact can be easily seen by noting the fact that as we set $\boldsymbol{\gamma}^{-2} \to 0$, we recover the LQR. Obviously, the H^∞ norm of the closed-loop system corresponding to an LQR cannot be ∞; so it is clear that as we set $\boldsymbol{\gamma}^2 \to \infty$, the closed-loop H^∞ norm does not approach ∞, and hence, γ does not equal the H^∞ norm in general. An implication of this fact is that the controller obtained via solving Eq. (36) with $\gamma^2 = \gamma_o^{\,2}$ is not one that minimizes the closed-loop H^2 norm subject to the H^∞ norm less than or equal to $\gamma_o^{\,2}$. For example, suppose the closed-loop H^∞ norm of the pure LQR control system is $\gamma_{LQR}^{\,2}$. Then, it is clear that this LQR is also the optimal solution to the minimum closed-loop H^2 norm control problem subject to any value of closed-loop H^∞ norm larger than $\gamma_{LQR}^{\,2}$, and this solution ought to be unique. But if we solve Eq. (36) with any finite $\gamma^2 > \gamma_{LQR}^{\,2}$, we will not obtain the LQR. This implies that the obtained solution cannot be optimal.

Other salient features of such an approach are: (i) the performance in consideration is usually (there are some exceptions such as [7]) over the infinite time horizon, (ii) the initial state is zero and, (iii) as already pointed out above, the performances in respect of the H^2 and H^∞ criteria are, loosely speaking, in a mutually-exclusive "either-or" situation; in that, it does not consider the *combined* effect when both the stochastic and the deterministic disturbances are present *simultaneously*. In other words, while the H^∞ norm gives a bound on the worst-case performance of the closed-loop system when there is a deterministic disturbance (but with no stochastic noises) and the H^2 norm gives indication on the performance in the presence of stochastic noise (but in the absence of deterministic disturbances), the combined effect on the closed-loop system when both types of exogenous disturbances are present is not explicitly and quantitatively addressed. Another problem we may identify is that the H^2 norm obtained in a mixed H^2/H^∞ design is actually in regards to the nominal model. When the system matrices are subject to perturbations, such H^2 norm would also be subject to change. Hence, the question of what would be the worst-case performance of the closed-loop system when the nominal model has been subject to perturbations remains to be answered. In other words, stability robustness may be established but nothing about the worst-case performance has been said, at least not quantitatively.

In this chapter, we revisit the mixed H^2/H^∞ control problem but from a different perspective. Specifically, we pose the original problem as a stochastic H^∞ control problem where the system is subject to both possibly "worst-case" (bounded) deterministic disturbances and stochastic "zero-mean" white noises *simultaneously*. We are interested in minimizing the worst-case *combined* effect of these disturbances, along with an uncertain nonzero initial condition [6, 11], on the closed-loop performance of the system. It should be clear that these problems would arise when one is interested to model and drive the state of a system from some uncertain nonzero value to the origin in some optimal way, in the presence of *both* high and low frequencies noises. Note that while we model the disturbances as two separate noises, they could actually be one single disturbance consisting of both low and high frequencies components, thus modeling a nonzero-mean disturbance.

Our results here will be derived from the Riccati differential equation related to the finite-horizon LQR problem, instead of like most existing approaches which are based on differential games theory. Rather than looking for a minimax saddle point as a solution, we minimize the expected worst-case cost and derive our solution by explicitly working with $\mathbf{P}(0)$ of the finite-horizon LQR Riccati equation, which is directly related to the value of the cost functional concerned when the initial state is nonzero. Clearly, if we assume the deterministic noise to be absent, the problem reduces to a stochastic LQR problem, and if we assume the stochastic white noise to be absent, we recover the standard H^∞ control problem. We believe that this approach offers us a new (in addition to the existing works such as [1–5], for example), and arguably also simpler, perspective to look at the solution to the H^∞ control problem, and H^∞ control with stochastic disturbances (compare [5, 8–10] for example).

It should be noted that the problem we consider here is different from the problem considered in some existing stochastic H^∞ control literature. For example, in [5, 10], the stochastic H^∞ approach was to incorporate stochastic disturbances into the system matrices to obtain a stochastic system with multiplicative white noise perturbations. This approach is quite different from the standard and familiar stochastic LQR/LQG approach, where the system matrices are assumed to be deterministic, not subjecting to stochastic perturbation or state-dependent multiplicative noises.

The organization of the chapter is as follows. In Section 2, we formally formulate the control problem of our interest. In Section 3, we state and derive some mathematical preliminaries. In Section 4, we consider how one may synthesize a finite-horizon controller with uncertain initial state. In Section 5, we consider steady-state solutions. Section 6 considers control with output feedback when the states are not directly measureable. In Section 7, we give an example. Section 8 contains our conclusions.

Notation: $||.||_2$ denotes Euclidean norm of the vector; $||.||_{L2}$ the L2-norm; $\underline{\Delta}$ denotes "equal by definition"; $E\{.\}$ denotes taking expectation; $(\int^T . \, dt)$ denotes the definite integral with respect to t from 0 to T; $||\mathbf{x}(t)||_T^2$ denotes $E(\int^T \mathbf{x}(t)^T\mathbf{x}(t)\, dt)$, and $A \geq B$ means A-B is positive semi-definite.

2. Problem formulation

Consider the continuous-time linear time-invariant system with input disturbances $\mathbf{w}(t)$ and $\mathbf{v}(t)$:

$$d\mathbf{x}(t)/dt = \mathbf{A}\mathbf{x}(t) + \mathbf{B_u}\mathbf{u}(t) + \mathbf{B_w}\mathbf{w}(t) + \mathbf{v}(t) \tag{2}$$

$$E\{\mathbf{x}(0)\} = \mathbf{0}, E\left\{\mathbf{x}(0)\mathbf{x}(0)^T\right\} \leq \mathbf{X}_0 \tag{3}$$

$\mathbf{A}$, $\mathbf{B_u}$, $\mathbf{B_w}$ are known real matrices, $\mathbf{u}(t)$ is the control input, $\mathbf{w}(t)$ is an unknown but bounded exogenous input with $E\{\int^T \mathbf{w}(t)^T\mathbf{w}(t)dt\} \leq 1$, $\mathbf{v}(t)$ is a stationary Gaussian zero-mean white noise with known covariance matrix $\mathbf{\Theta}$. For simplicity, we assume that the system is controllable and observable in the case of output feedback.

The feedback control problem of our interest is to find a linear causal control law $\mathbf{u}(t) = \mathbf{K}(t)\mathbf{x}(t)$ to *minimize*

$$J_{\mathbf{u}} = \left\{||\boldsymbol{\Pi}^{1/2}\mathbf{x}(t)||_T^{\ 2} + ||\mathbf{R}^{1/2}\mathbf{u}(t)||_T^{\ 2}\right\} + E\{\mathbf{x}^T{}_T\mathbf{Q}\ \mathbf{x}_T\} \tag{4}$$

where $\mathbf{\Pi}$ and $\mathbf{R}$ are positive-definite and $\mathbf{Q}$ is positive-semi-definite weight matrices.

Note that (4) may be re-written as

$$J_{\mathbf{u}} = ||\boldsymbol{\Pi}^{1/2}\mathbf{x}(t)||_T^{\ 2} + ||\mathbf{R}^{1/2}\mathbf{K}(t)\mathbf{x}(t)||_T^{\ 2} + E||\mathbf{Q}^{1/2}\mathbf{x}(T)||_2 \tag{5}$$

In what follows, we shall sometimes drop the argument t (especially for the system matrices), or denote $\mathbf{z}(t)$ as $\mathbf{z}_t$, for notational simplicity when there is no danger of confusion.

3. Preliminaries

Lemma 3.1: Let $\mathbf{A}$, $\mathbf{B}$, $\mathbf{L} = \mathbf{L}^T$, and $\mathbf{Q_L} = \mathbf{Q_L}^T$ be given matrices. Suppose there exits on the interval $0 \leq t \leq T$ a solution $\mathbf{P_o}(t)$ of the differential Riccati equation (DRE)

$$\begin{aligned} d\mathbf{P}(t)/dt &= -\mathbf{P}(t)\mathbf{A}-\mathbf{A}^T\mathbf{P}(t) + \mathbf{P}(t)\mathbf{B}\mathbf{B}^T\mathbf{P}(t)-\mathbf{L} \\ \mathbf{P}(T) &= \mathbf{Q_L}. \end{aligned} \tag{6}$$

Then the minimum of the cost functional

$$J\Delta\int^T (\mathbf{x}^T\mathbf{L}\mathbf{x} + \mathbf{w}^T\mathbf{w})dt + \mathbf{x}^T(T)\mathbf{Q_L}\mathbf{x}(T) \tag{7}$$

for the system $d\mathbf{x}/dt = \mathbf{A}\mathbf{x}(t) + \mathbf{B}\mathbf{w}(t) + \mathbf{v}$, $E\{\mathbf{v}(t)\} = \mathbf{0}$, $E\{\mathbf{v}(t)\mathbf{v}(t)^T\} = \boldsymbol{\Theta}$, $E\{\mathbf{x}(0)\} = \mathbf{0}$, and $E\{\mathbf{x}(0)\mathbf{x}(0)^T\} = \mathbf{X}_0$ is given by

$$J_{min} = \mathrm{Tr}\left\{\mathbf{P_o}(0)\mathbf{X}_0 + \int^T \mathbf{P_o}(t)\boldsymbol{\Theta}dt\right\} \tag{8}$$

The minimizing $\mathbf{w}(t)$ is given by

$$\mathbf{w}(t) = -\mathbf{B}^T\mathbf{P_o}(t)\mathbf{x}(t) \tag{9}$$

Proof: Well-known stochastic LQR result (see Ref. [12], p. 221).

We next give a similar result concerning the deterministic problem and the related Riccati differential equation, which we will find useful later in studying some of the properties of RDE (6).

Lemma 3.2: Let $\mathbf{A}(t)$, $\mathbf{B}(t)$, $\mathbf{L}(t) = \mathbf{L}(t)^T$, and $\mathbf{Q_L} = \mathbf{Q_L}^T$ be given matrices. Suppose there exists on the interval $0 \leq t \leq T$ a solution $\mathbf{P_o}(t)$ of the differential Riccati equation (DRE)

$$\begin{aligned} d\mathbf{P}(t)/dt &= -\mathbf{P}(t)\mathbf{A}-\mathbf{A}^T\mathbf{P}(t) + \mathbf{P}(t)\mathbf{B}\mathbf{B}^T\mathbf{P}(t)-\mathbf{L} \\ \mathbf{P}(T) &= \mathbf{Q_L}. \end{aligned} \tag{10}$$

Then the cost functional

$$\eta \underline{\Delta} \int^{T} \left(\mathbf{x}^{T}\mathbf{L}\mathbf{x} + \mathbf{w}^{T}\mathbf{w}\right)dt + \mathbf{x}^{T}(T)\mathbf{Q}_{L}\mathbf{x}(T) \tag{11}$$

for system $d\mathbf{x}/dt = \mathbf{A}\mathbf{x}(t) + \mathbf{B}\mathbf{w}(t)$: $\mathbf{x}(0) = \mathbf{x}_0$ is given by

$$\eta = \left|\left|\mathbf{w}(t) + \mathbf{B}^{T}\mathbf{P}_{o}(t)\mathbf{x}(t)\right|\right|_{T}^{2} + \mathbf{x}_0{}^{T}\mathbf{P}_{o}(0)\mathbf{x}_0. \tag{12}$$

The minimum value of η is given by $\mathbf{x}_0{}^{T}\mathbf{P}_0(0)\mathbf{x}_0$, and the minimizing $\mathbf{w}(t)$ is given by

$$\mathbf{w}(t) = -\mathbf{B}^{T}\mathbf{P}_{o}(t)\mathbf{x}(t) \tag{13}$$

Proof: See Ref. [13], (proof Theorem 1, p. 131.)

Remark 3.1: Note that $\mathbf{Q}_L$ and $\mathbf{L}$ are symmetric but not necessarily nonnegative definite. Furthermore, as pointed out in [13], because $\mathbf{L}$ is not assumed to be nonnegative definite, it is not generally true that a minimum η exists. However, we know from LQR theory that such minimum η exists when $\mathbf{L}$ is nonnegative definite. Hence, it is reasonable to expect that a minimum η exists if $\mathbf{L}$ is not "exceedingly" negative definite.

Lemma 3.3: Let $\mathbf{A}$, $\mathbf{B}$, $\mathbf{L} = \mathbf{L}^{T} \leq \mathbf{0}$, and $\mathbf{Q}_L = \mathbf{Q}_L{}^{T}$ be given matrices. Suppose there exits on the interval $0 \leq t \leq T$ a solution $\mathbf{P}_s(t)$ of the differential Riccati inequality (DRI), [14]:

$$\begin{aligned} d\mathbf{P}(t)/dt &\geq -\mathbf{P}(t)\mathbf{A} - \mathbf{A}^{T}\mathbf{P}(t) + \mathbf{P}(t)\mathbf{B}\mathbf{B}^{T}\mathbf{P}(t) - \mathbf{L} \\ \mathbf{P}(T) &= \mathbf{Q}_{L}. \end{aligned} \tag{14}$$

Then for the system $d\mathbf{x}/dt = \mathbf{A}\mathbf{x}(t) + \mathbf{B}\mathbf{u}(t)$: $\mathbf{x}(0) = \mathbf{x}_0$, η as defined in (11) satisfies:

$$\eta \geq \left|\left|\mathbf{w}(t) + \mathbf{B}^{T}\mathbf{P}_{s}(t)\mathbf{x}(t)\right|\right|_{T}^{2} + \mathbf{x}_0{}^{T}\mathbf{P}_{s}(0)\mathbf{x}_0. \tag{15}$$

Proof: Since $\mathbf{L} \leq \mathbf{0}$, there exists $\mathbf{L}_1 \leq \mathbf{L}$ such that

$$\begin{aligned} d\mathbf{P}_{s}(t)/dt &= -\mathbf{P}_{s}(t)\mathbf{A} - \mathbf{A}^{T}\mathbf{P}_{s}(t) + \mathbf{P}_{s}(t)\mathbf{B}\mathbf{B}^{T}\mathbf{P}_{s}(t) - \mathbf{L}_1 \\ \mathbf{P}_{s}(T) &= \mathbf{Q}_{L}. \end{aligned} \tag{16}$$

Let

$$\eta_1 \underline{\Delta} \int^{T} \left(\mathbf{x}^{T}\mathbf{L}_1\mathbf{x} + \mathbf{w}^{T}\mathbf{w}\right)dt + \mathbf{x}^{T}(T)\mathbf{Q}_{L}\mathbf{x}(T) \tag{17}$$

Clearly, $\eta \geq \eta_1$ for any $\mathbf{x}$ and $\mathbf{w}$ since $\mathbf{L} \geq \mathbf{L}_1$. Applying Lemma 3.2 with $\mathbf{L}$ replaced with $\mathbf{L}_1$ gives $\eta_1 = ||\mathbf{w}(t) + \mathbf{B}^{T}\mathbf{P}_s(t)\mathbf{x}(t)||_T{}^2 + \mathbf{x}_0{}^{T}\mathbf{P}_s(0)\mathbf{x}_0$ and completes the proof.

Remark 3.2: Since the inequality sign in (14) is non-strict, the set of solutions that solve Lemma 3.3 contains $\mathbf{P}_0(t)$ of Lemma 3.2 as a member.

Remark 3.3: In Lemmas 3.1-3.3, $\mathbf{P}(t)$ is *independent* of the initial condition $\mathbf{x}_0$, even though the cost η is a function of $\mathbf{x}_0$.

Proposition 3.1: Let $\mathbf{P}_0(t)$ be the solution that satisfies Lemma 3.2 and $\mathbf{P}_1(t)$ be any solution that satisfies (14). Then

$$\mathbf{P_o}(0) \geq \mathbf{P}_1(0) \tag{18}$$

Proof: Since $\eta \geq \eta_1$ for any $\mathbf{x}$ and $\mathbf{w}$, it follows that $\mathbf{x}_0{}^T\mathbf{P_o}(0)\mathbf{x}_0 \geq \mathbf{x}_0{}^T\mathbf{P}_1(0)\mathbf{x}_0$ for any $\mathbf{x}_0$. Thus (18) must follow.

Remark 3.4: Note time 0 may be any time before time T. So (18) also implies $\mathbf{P_o}(t) \geq \mathbf{P}_1(t)$, all $t < T$.

Proposition 3.2: Let $\mathbf{P}_1(t)$ be a solution and $\mathbf{P}_2(t)$ be any solution that satisfies (14). Then $\mathbf{P}_1(t)$ also satisfies Lemma 3.2 if and only if

$$\mathrm{Tr}[\mathbf{P}_1(0)] \geq \mathrm{Tr}[\mathbf{P}_2(0)] \tag{19}$$

for all admissible $\mathbf{P}_2(t)$.

Proof: Necessity is obvious from Proposition 3.1. To prove sufficiency, suppose there is a $\mathbf{P}_1(t)$ that satisfies (19) but not a solution to Lemma 3.2. Since $\mathbf{P}_1(t)$ is a solution to (14), it follows from Proposition 3.1 that $\mathbf{P_o}(0) \geq \mathbf{P}_1(0)$. But $\mathbf{P_o}(0)$ is also an admissible solution of Lemma 3.3. Then, since $\mathrm{Tr}[\mathbf{P}_1(0)] \geq \mathrm{Tr}[\mathbf{P_o}(0)]$ but $\mathbf{P}_1(0) \neq \mathbf{P_o}(0)$, it follows that $\mathbf{P_o}(0) \geq \mathbf{P}_1(0)$ does not hold. This contradicts Proposition 3.1 and completes the proof.

Proposition 3.3: Let $\mathbf{P}_1(t)$ be a solution to Lemma 3.2 and $\mathbf{P}_2(t)$ be any solution that satisfies (14). Then the following statements are equivalent:

$$\begin{gathered} \mathrm{Tr}\{\mathbf{P}_1(0)\} \geq \mathrm{Tr}\{\mathbf{P}_2(0)\} \\ d\{\mathrm{Tr}\mathbf{P}_1(t)]/dt \leq d\{\mathrm{Tr}\mathbf{P}_2(t)\}/dt \end{gathered} \tag{20}$$

Proof: Obvious consequence of Bellman's principle of optimality, since both $\mathbf{P}_1(t)$ and $\mathbf{P}_2(t)$ are solutions computed backward in time from final time T with $\mathbf{P}_s(T) = \mathbf{Q_L}$.

Remark 3.4: Proposition 3.3 tells us that in order to find the minimizing $\mathbf{w}(t)$ which minimizes (7) and (11) from the set of sub-optimal solutions characterized by Lemma 3.3, it suffices to search for one with the maximum $\mathrm{Tr}\mathbf{P}(t)$ or minimum $d\{\mathrm{Tr}\mathbf{P}(t)\}/dt$ over all $t \in [0,T]$. These properties will be useful later when we search for the optimal controller via linear matrix inequality (LMI) as follow.

Consider the LMI:

$$\begin{bmatrix} \mathbf{\Gamma}(t) + \mathbf{PA} + \mathbf{A}^T\mathbf{P} - \mathbf{L} & \mathbf{PB} \\ \mathbf{B}^T\mathbf{P} & \mathbf{I} \end{bmatrix} \geq \mathbf{0} \tag{21}$$

Applying Schur transformation to (21) then gives us $\mathbf{\Gamma}(t) \geq -\mathbf{P}(t)\mathbf{A} - \mathbf{A}^T\mathbf{P}(t) + \mathbf{P}(t)\mathbf{B}\mathbf{B}^T\mathbf{P}(t) - \mathbf{L}$. It follows from Proposition 3.3 that the $\mathbf{\Gamma}(t)$ that solves $\min\mathrm{Tr}\mathbf{\Gamma}$ subject to (21) at all $t \in [0,T]$ will give us the desired $d\mathbf{P}/dt$ that solves $d\mathbf{P}/dt = -\mathbf{PA} - \mathbf{A}^T\mathbf{P} + \mathbf{PBB}^T\mathbf{P}(t) - \mathbf{L}$ for all $t \in [0,T]$, where $\mathbf{\Gamma} = d\mathbf{P}/dt$.

Before proceeding further, let us recollect what we have done so far. Lemmas 3.1 and 3.2 show that the minimizing $\mathbf{w}(t)$ is the same whether the system is subjected to an input stochastic white noise or not. However, the minimum costs resulted will be different as given by (8) and $\mathbf{x}_0{}^T\mathbf{P_o}(0)\mathbf{x}_0$ (implicitly given by (12)), respectively. Then from Lemma 3.3 to Proposition 3.3, we transformed the problem of finding $\mathbf{P}(t)$ for (6) into an equivalent LMI problem.

4. Finite-horizon state regulation

Let $\mathbf{u} = \mathbf{Kx}$ where $\mathbf{K}$ is the state feedback controller. Let $\mathbf{A_{cl}} = \mathbf{A} + \mathbf{B_u K}$. Let

$$\mathbf{L} = -\left(\mathbf{\Pi} + \mathbf{K}^{\mathrm{T}}\mathbf{R}\mathbf{K}\right) \tag{22}$$

Then it is clear

$$\mathrm{J}_{\mathbf{u}} = \int^{\mathrm{T}} \mathbf{x}^{\mathrm{T}}\left(\mathbf{\Pi} + \mathbf{K}^{\mathrm{T}}\mathbf{R}\mathbf{K}\right)\mathbf{x}\,\mathrm{dt} + \mathbf{x}(\mathrm{T})^{\mathrm{T}}\mathbf{Q}\mathbf{x}(\mathrm{T}) \tag{23}$$

We are interested in finding a $\mathbf{w}(\mathrm{t})$ to maximize $\mathrm{J}_{\mathbf{u}}$. For this purpose, we consider

$$\left\{-\gamma^{-2}\int^{\mathrm{T}} \mathbf{x}^{\mathrm{T}}\left(\mathbf{\Pi} + \mathbf{K}^{\mathrm{T}}\mathbf{R}\mathbf{K}\right)\mathbf{x}\right)\mathrm{dt} + \mathbf{x}(\mathrm{T})^{\mathrm{T}}\mathbf{Q}\mathbf{x}(\mathrm{T})\right\} + \|\mathbf{w}\|_{\mathrm{T}}{}^{2} - \|\mathbf{w}\|_{\mathrm{T}}{}^{2} \tag{24}$$

Let

$$\begin{aligned} \mathrm{d}\mathbf{P}(\mathrm{t})/\mathrm{dt} &= -\mathbf{P}(\mathrm{t})\mathbf{A}_{\mathbf{cl}} - \mathbf{A}_{\mathbf{cl}}{}^{\mathrm{T}}\mathbf{P}(\mathrm{t}) + \mathbf{P}(\mathrm{t})\mathbf{B}\mathbf{B}^{\mathrm{T}}\mathbf{P}(\mathrm{t}) + \gamma^{-2}\left(\mathbf{\Pi} + \mathbf{K}^{\mathrm{T}}\mathbf{R}\mathbf{K}\right) \\ \mathbf{P}(\mathrm{T}) &= -\gamma^{-2}\mathbf{Q}. \end{aligned} \tag{25}$$

Lemma 4.1: Let $\mathbf{A}_{\mathbf{cl}} = \mathbf{A} + \mathbf{B}_{\mathbf{u}}\mathbf{K}$ and $\mathbf{P}_{\mathbf{o}}$ be given by (25) for some chosen value of γ^2. Then for the system (2), the state feedback control law $\mathbf{u} = \mathbf{K}\mathbf{x}$ achieves the close loop performance for all admissible $\mathbf{w}(\mathrm{t})$:

$$\mathrm{J}_{\mathbf{u}} \le \gamma^2\left(\|\mathbf{w}\|_{\mathrm{T}}{}^{2} - \mathrm{Tr}\left\{\mathbf{P}_{\mathbf{o}}(0)\mathbf{X}_0 - \int^{\mathrm{T}} \mathbf{P}_{\mathbf{o}}(\mathrm{t})\mathbf{\Theta}\mathrm{dt}\right\}\right) \tag{26}$$

And the worst case $\mathbf{w}_{\mathbf{o}}$ achieves

$$\mathrm{J}_{\mathbf{u}} = \gamma^2\left(\|\mathbf{w}_{\mathbf{o}}\|_{\mathrm{T}}{}^{2} - \mathrm{Tr}\left\{\mathbf{P}_{\mathbf{o}}(0)\mathbf{X}_0 - \int^{\mathrm{T}} \mathbf{P}_{\mathbf{o}}(\mathrm{t})\mathbf{\Theta}\mathrm{dt}\right\}\right) \tag{27}$$

Proof: From Lemma 3.1,

$$\begin{aligned} &\min{}_{\mathbf{w}}\left\{-\gamma^{-2}\int^{\mathrm{T}} \mathbf{x}^{\mathrm{T}}\left(\mathbf{\Pi} + \mathbf{K}^{\mathrm{T}}\mathbf{R}\mathbf{K}\right)\mathbf{x}\right)\mathrm{dt} + \|\mathbf{w}\|_{\mathrm{T}}{}^{2}\right\} - \|\mathbf{w}\|_{\mathrm{T}}{}^{2} \\ &= \mathrm{Tr}\left\{\mathbf{P}_{\mathbf{o}}(0)\mathbf{X}_0 + \int^{\mathrm{T}} \mathbf{P}_{\mathbf{o}}(\mathrm{t})\mathbf{\Theta}\mathrm{dt}\right\} - \|\mathbf{w}\|_{\mathrm{T}}{}^{2} \end{aligned}$$

Multiply throughout by $-\gamma^2$ we obtain:

$$\begin{aligned} &\max{}_{\mathbf{w}}\left\{\mathrm{J}_{\mathrm{u}} - \gamma^2\|\mathbf{w}\|_{\mathrm{T}}{}^{2}\right\} + \gamma^2\|\mathbf{w}\|_{\mathrm{T}}{}^{2} \\ &= -\gamma^2\left\{\mathrm{Tr}\left\{\mathbf{P}_{\mathbf{o}}(0)\mathbf{X}_0 + \int^{\mathrm{T}} \mathbf{P}_{\mathbf{o}}(\mathrm{t})\mathbf{\Theta}\mathrm{dt}\right\} + \|\mathbf{w}\|_{\mathrm{T}}{}^{2}\right\} \end{aligned} \tag{28}$$

This completes the proof.

Remark 4.2: From (26), we recover the conventional H^{∞} inequality [4, 12]:

$$\mathrm{J}_{\mathbf{u}} \le \gamma^2\|\mathbf{w}\|_{\mathrm{T}}{}^{2} \text{if } \mathbf{x}_0 = \mathbf{0} \text{ and } \mathbf{\Theta} = \mathbf{0} \tag{29}$$

Corollary 4.1: Let $\mathbf{A_{cl}} = \mathbf{A} + \mathbf{B_u K}$ and $\mathbf{P_0}$ be given by Lemma 4.1. If $\mathbf{w_0} = -\mathbf{B_w}^T\mathbf{P_0}(t)\mathbf{x}(t)$, then

$$J_{\mathbf{u}} = \gamma^2\left(\|\mathbf{w_o}\|_T{}^2 - \mathrm{Tr}\left\{\mathbf{P_o}(0)X_0 - \int^T \mathbf{P_o}(t)\mathbf{\Theta}dt\right\}\right. \tag{30}$$

Proof: Obvious from (28) and Lemma 3.1. This completes the proof.

To derive the optimal controller for Lemma 4.1, We substitute $A_{cl} = [A + B_uK]$ into (6) with $P(T) = -\gamma^{-2}Q$. Thus we obtain

$$d\mathbf{P}(t)/dt + \mathbf{P}(t)[\mathbf{A} + \mathbf{B_u K}] + [\mathbf{A} + \mathbf{B_u K}]^T\mathbf{P}(t) - \mathbf{P}(t)\mathbf{B_w B_w}^T\mathbf{P}(t) - \gamma^{-2}(\mathbf{\Pi} + \mathbf{K}^T\mathbf{RK}) = \mathbf{0} \tag{31}$$

From (27), it is clear that the "larger" the $\mathrm{Tr}\mathbf{P_0}(0)\mathbf{X}_0$ and $\mathrm{Tr}\int^T\mathbf{P_0}(t)\mathbf{\Theta}dt$, the smaller J_u would be. This implies that we want to find the least negative-definite $\mathbf{P_0}(t)$ for all $t < T$. Proposition 3.2 implies that we should search for the $\mathbf{P_0}(t)$ with the least negative $\mathrm{Tr}\mathbf{P}(t)$. Also, since we will be working backward in time when solving (31), this implies that we want to find a $\mathbf{K}(t)$ which minimizes $\mathrm{Tr}\{d\mathbf{P}/dt\}$ at every instant of t (principle of optimality), to maximize $\mathrm{Tr}\mathbf{P_0}(t)$ (Proposition 3.1 and 3.2). To achieve this, we note

$$d\mathbf{P}(t)/dt = -\mathbf{P}(t)[\mathbf{A} + \mathbf{B_u K}] - [\mathbf{A} + \mathbf{B_u K}]^T\mathbf{P}(t) + \mathbf{P}(t)\mathbf{B_w B_w}^T\mathbf{P}(t) + \gamma^{-2}(\mathbf{\Pi} + \mathbf{K}^T\mathbf{RK}) \tag{32}$$

Differentiate $\mathrm{Tr}\{d\mathbf{P}/dt\}$ with respect to $\mathbf{K}$ and set to $\mathbf{0}$:

$$\mathbf{0} = -2\mathbf{P}(t)\mathbf{B_u} + 2\gamma^{-2}\mathbf{K}^T\mathbf{R} \tag{33}$$

$$\begin{gathered}\mathbf{K} = \gamma^2\mathbf{R}^{-1}\mathbf{B_u}^T\mathbf{P}\\ d^2\mathrm{Tr}\{d\mathbf{P}/dt\}/d\mathbf{K}^2 = 2\gamma^{-2}\mathbf{R} > \mathbf{0}\end{gathered} \tag{34}$$

Substituting (34) into (32), we obtain

$$d\mathbf{P}/dt = -\mathbf{PA} - \mathbf{A}^T\mathbf{P} + \mathbf{P}(\mathbf{B_w B_w}^T - \gamma^2\mathbf{B_u R}^{-1}\mathbf{B_u}^T)\mathbf{P} + \gamma^{-2}\mathbf{\Pi} \tag{35}$$

If we let $\gamma^{-2}\mathbf{P}_\gamma = -\mathbf{P}$, then we may show that (35) is equivalent to

$$-d\mathbf{P}_\gamma/dt = \mathbf{P}_\gamma\mathbf{A} + \mathbf{A}^T\mathbf{P}_\gamma - \mathbf{P}_\gamma(\mathbf{B_u R}^{-1}\mathbf{B_u}^T - \gamma^{-2}\mathbf{B_w B_w}^T)\mathbf{P}_\gamma + \mathbf{\Pi} \tag{36}$$

Note that (36) is the same RDE obtained in the conventional approach to H^∞ control with $\mathbf{R} = \mathbf{I}$ and trivial initial state, ([4, 12]). We have, thus, established the connection between (35) and (36). However in our derivation here, we are able to concurrently derive a quantitative cost corresponding to the solution based on the initial condition and the simultaneous disturbances, as is in the case for stochastic LQR control.

Theorem 4.1: Let $\mathbf{A_{cl}} = \mathbf{A} + \mathbf{B_u K}$ and $\mathbf{P_0}$ be given by (35) with $P(\mathbf{T}) = -\gamma^{-2}Q$. Then $\mathbf{K} = \gamma^2\mathbf{R}^{-1}\mathbf{B_u}^T\mathbf{P}$ achieves

$$J_{\mathbf{u}} \le \gamma^2\left(\|\mathbf{w}\|_T{}^2 - \mathrm{Tr}\left\{\mathbf{P_o}(0)\mathbf{X}_0 - \int^T \mathbf{P_o}(t)\mathbf{\Theta}dt\right\}\right. \tag{37}$$

And the worst case $\mathbf{w_0}$ achieves

$$J_u = \gamma^2\left(\|w\|_T{}^2 - \mathrm{Tr}\left\{ P_o(0)X_0 - \int^T P_o(t)\Theta dt \right\}\right) \tag{38}$$

Proof: Substituting $K = \gamma^2 R^{-1} B_u{}^T P$ into (25) and applying Lemma 4.1 completes the proof.

To compute $\|\mathbf{w_0}\|_T$, the system with $\mathbf{w_0} = -\mathbf{B_w}^T\mathbf{P_0}(t)\mathbf{x}(t)$ is given by

$$\begin{aligned} d\mathbf{x}/dt &= \mathbf{Ax} + \mathbf{B_u Kx} - \mathbf{B_w B_w}^T\mathbf{P_o x} + \mathbf{v} \\ &= \left(\mathbf{A} + \mathbf{B_u K} - \mathbf{B_w B_w}^T\mathbf{P_o}\right)\mathbf{x} + \mathbf{v} \\ &= \mathbf{A_w x} + \mathbf{v} \end{aligned} \tag{39}$$

Define

$$E\left\{\mathbf{x}(t)\mathbf{x}(t)^T\right\} = \mathbf{X}_t \tag{40}$$

Then $\mathbf{X}_t$ satisfies [12]:

$$d\mathbf{X}_t/dt = \mathbf{A_w X}_t + \mathbf{X}_t\mathbf{A_w}^T + \mathbf{\Theta}, \mathbf{X}_t(0) = \mathbf{X}_0 \tag{41}$$

$$\begin{aligned} E\|\mathbf{w_o}\|_T{}^2 &= E\left\{\int^T \mathbf{x}(t)^T\mathbf{P_o B_w B_w}^T\mathbf{P_o x}(t)dt\right\} \\ &= \mathrm{Tr}\left[\int^T \mathbf{P_o B_w B_w}^T\mathbf{P_o X}_t dt\right] \end{aligned} \tag{42}$$

To check if the value of γ^2 selected is the correct one, we require γ^2 to be such that $\|\mathbf{w_0}\|_T{}^2 = 1$. This could be achieved by varying and searching over values of γ^2. We note that $\|\mathbf{w_o}\|_T{}^2$ is generally a decreasing function of γ^2, as is already known from conventional H^∞ control theory.

Once the desired value of γ^2 and hence $P(t)$, $0 \le t \le T$, are found, we may then compute the cost via Theorem 4.1. We summarize the results in the following conceptual algorithm:

Conceptual Algorithm 1.

Step 1: Choose a value of γ^2 and compute $\mathbf{P}(t)$ with (35). If $\mathbf{P}(t)$ does not exist over the entire time interval $[0, T]$, the chosen γ^2 is too small. Set lower bound $\gamma_l{}^2 = \gamma^2$, increase γ^2, and repeat this step.

Step 2: If $\mathbf{P}(t)$ exists over the entire time interval $[0, T]$, compute $\|\mathbf{w_0}\|_T{}^2$. If $\|\mathbf{w_0}\|_T{}^2 = 1$, we have the desired solution. The optimal state control law is given by (34).

Step 3: If $\|\mathbf{w_0}\|_T{}^2 > 1$, the chosen γ^2 is too small. Set lower bound $\gamma_l{}^2 = \gamma^2$, increase γ^2, and go to step 1.

Step 4: If $\|\mathbf{w_0}\|_T{}^2 < 1$, the chosen γ^2 is too large. Set upper bound $\gamma_u{}^2 = \gamma^2$, decrease γ^2, and go to step 1.

Remark 4.1 It is important to note the presence of $\mathbf{\Theta}$ in (41). In other words, $\mathbf{v}$ does have an impact on the value of $\mathbf{X}_t$ and hence $\mathbf{w_0}$. This distinguishes our approach here from the previous mixed H^2/H^∞ solutions which do not explicitly take $\mathbf{\Theta}$ into consideration.

5. Infinite-horizon state regulation

It is well known from conventional H^∞ control theory [4, 12, 15] that if γ^2 is large enough, the feedback gain would approach a steady-state value $\mathbf{K}_\infty$ far from the final time. In such a case, the feedback system would approach a linear time-invariant system. Unfortunately, this is *not* the steady-state solution for us, at least not for the problem formulated in Section 2. This is because as $T \to \infty$ and as system approaches steady-state, $\lim_{T \to \infty}\{\mathrm{Tr}\int^T \mathbf{\Theta P_o}(t)dt\} \to \lim_{T \to \infty}\mathrm{Tr}\{T\mathbf{\Theta P_o}(\infty)\} \to -\infty$ will dominate the other two cost components, namely $\mathrm{Tr}\mathbf{P_0}(\infty)\mathbf{X}_0$ and $\lim_{T \to \infty}||\mathbf{w}||_T = ||\mathbf{w}||_{L2} < \infty$. It follows that for formulations which require $||\mathbf{w}||_{L2} \le \infty$ and finite $\mathbf{x}_0$, the optimal infinite-horizon solution is indeed the LQR since it minimizes $\mathrm{Tr}\{\mathbf{\Theta P_o}(\infty)\}$ with $\gamma^{-2} = 0$.

Instead, we shall explore an alternative formulation. First, we note that there exists a steady-state solution $\mathbf{P}_\infty$ to the algebraic Riccati equation means $\mathbf{P}_\infty$ is the solution to the RDE with $-\gamma^{-2}\mathbf{Q} = \mathbf{P}(t) = \mathbf{P}_\infty$, $0 \le t \le T$. Substituting this into the RHS of (26), we obtain and define

$$\eta_{ss} \le \gamma^2 \left(E\{\|\mathbf{w}\|_T^2\} - \mathrm{Tr}\{\mathbf{P}_\infty \mathbf{X}_0 + T\mathbf{\Theta}\mathbf{P}_\infty)\}\right) \tag{43}$$

where

$$\eta_{ss} \underline{\Delta} E\left\{||\mathbf{\Pi}^{1/2}\mathbf{x}||_T^2 + ||\mathbf{R}^{1/2}\mathbf{u}||_T^2 - \gamma^2\mathbf{x}(T)^T\mathbf{P}_\infty\mathbf{x}(T)\right\} \tag{44}$$

Dividing (43) by T gives

$$\eta_{ss}/T \le \gamma^2 \left(E\{\|\mathbf{w}\|_T^2/T\} - \mathrm{Tr}\mathbf{P}_\infty \mathbf{X}_0/T - \mathrm{Tr}\{\mathbf{\Theta}\mathbf{P}_\infty)\}\right) \tag{45}$$

Assume $\lim_{T \to \infty}E\{||\mathbf{w}||_T^2\}/T = p^2$. Note that p^2 may be interpreted as "power". We obtain the following result:

Lemma 5.1: Let $\mathbf{P}_\infty$, $\mathbf{K}_\infty$, and $\mathbf{A_{cl}} = \mathbf{A} + \mathbf{B_u K}_\infty$ be the steady-state solutions corresponding to (35) with $T \to \infty$. If $\mathbf{A} + \mathbf{B_u K}_\infty - \mathbf{B_w B_w}^T\mathbf{P}_\infty$ is stable, then $\mathbf{K}_\infty$ achieves the infinite-horizon performance

$$\eta_\infty \underline{\Delta} \lim_{T\to\infty}(\eta_{ss}/T) \le \gamma^2 \left(p^2 - \mathrm{Tr}\{\mathbf{\Theta}\mathbf{P}_\infty\}\right) \tag{46}$$

Proof: If $\mathbf{A} + \mathbf{B_u K}_\infty - \mathbf{B_w B_w}^T\mathbf{P}_\infty$ is stable, it implies that if we let $\mathbf{Q} = -\gamma^{-2}\mathbf{P}_\infty$, then $\mathbf{P}(t) = \mathbf{P}_\infty$ exists for all $t < \infty$. It follows that (45) holds. Noting $\lim_{T \to \infty}\{\mathbf{P}_\infty\mathbf{X}_0/T\} = \mathbf{0}$ completes the proof.

Remark 5.1: Substituting $\lim_{T \to \infty}E\{||\mathbf{w}||_T^2\}/T = p^2$ into and setting $\mathbf{\Theta} = \mathbf{0}$ in (46) shows that the H^∞ norm of the closed-loop system from $\mathbf{w}$ to η_{ss} is bounded by γ. Similarly, setting $\mathbf{w} = \mathbf{0}$ shows that the (from map) steady-state LQR cost averaged over T as $T \to \infty$ would be $\gamma^2\mathrm{Tr}\{-\mathbf{\Theta P}_\infty\}$.

Finally, since $\mathbf{A} + \mathbf{B_u K} - \mathbf{B_w B_w}^T\mathbf{P}_\infty$ is stable, there exists a $\mathbf{X}_\infty$ which satisfies the following Lyapunov equation:

$$\left[\mathbf{A} + \mathbf{B_u K} - \mathbf{B_w B_w}^T\mathbf{P}_\infty\right]\mathbf{X}_\infty + \mathbf{X}_\infty\left[\mathbf{A} + \mathbf{B_u K} - \mathbf{B_w B_w}^T\mathbf{P}_\infty\right] + \mathbf{\Theta} = \mathbf{0} \tag{47}$$

So $\mathbf{w_0} = \mathbf{B_w}^T\mathbf{P}_\infty\mathbf{x}$ gives

$$\begin{aligned} E\{\|\mathbf{w_o}\|_T^2\} &= \int^T \mathrm{Tr}\left[\mathbf{B_w}^T\mathbf{P}_\infty\right]^T\left[\mathbf{B_w}^T\mathbf{P}_\infty\right]\mathbf{X}_\infty dt \\ &= \mathrm{Tr}\left[\mathbf{B_w}^T\mathbf{P}_\infty\right]^T\left[\mathbf{B_w}^T\mathbf{P}_\infty\right]\mathbf{X}_\infty T \end{aligned} \tag{48}$$

since $\mathbf{B_w}$, $\mathbf{P}_\infty$, and $\mathbf{X}_\infty$ are all constant matrices. Dividing both sides of (48) by T gives

$$\lim_{T\to\infty} E\{\|\mathbf{w_o}\|_T^2/T\} = \mathrm{Tr}\left[\mathbf{B_w}^T\mathbf{P}_\infty\right]^T\left[\mathbf{B_w}^T\mathbf{P}_\infty\right]\mathbf{X}_\infty. \tag{49}$$

where $\mathbf{X}_\infty$ is given by (47).

We may further assume that $\lim_{t\to\infty} E\{\|\mathbf{w}\|_2^2\} = \lim_{T\to\infty}\|\mathbf{w}\|_T^2/T \le 1$, and $\mathbf{\Theta}$ is unknown with worst-case $\mathrm{Tr}\mathbf{\Theta} \le \sigma$ where $\sigma \ge 0$. This would correspond to the system being subject to both worst case $\mathbf{w}(t)$ and $\mathbf{v}(t)$, where $\mathbf{v}(t)$ belongs to the class of white noises that satisfies $E\{\mathbf{v}^T\mathbf{v}\} \le \sigma$. The controller designed may then be interpreted as one which optimizes expected worst-case performance in the presence of both bounded-power deterministic disturbances and stochastic white noises. We summarize the procedure of how to find such a controller in conceptual algorithm 2.

Conceptual Algorithm 2.

Step 1: Guess a value of γ^2 and compute $\mathbf{P}_\infty$ from (35) with $d\mathbf{P}/dt = \mathbf{0}$. If $\mathbf{P}_\infty < 0$ does not exist, the chosen γ^2 is too small. Set lower bound $\gamma_l^2 = \gamma^2$, increase γ^2, and repeat this step.

Step 2: If P_∞ exists, find Θ_o which solves $\min_{\mathrm{Tr}\Theta=\sigma}\{\mathrm{Tr}P_\infty\Theta\}$. Compute X_∞ from (47) with $\Theta = \Theta_o$ and compute $p^2 = \lim_{t\to\infty}\|w_o\|_2$ via (48). If $p^2 = 1$, we have the desired solution.

Step 3: $p^2 > 1$, the chosen γ^2 is too small. Set lower bound $\gamma_l^2 = \gamma^2$, increase γ^2, and go to step 1.

Step 4: $p^2 < 1$, the chosen γ^2 is too large. Set upper bound $\gamma_u^2 = \gamma^2$, increase γ^2, and go to step 1.

6. H^∞ control with output feedback

In this section, we consider the case where the state of the system is not available for feedback. In this case, we consider dynamical controllers of the form $\mathbf{u}(t) = \mathbf{K}(t)*\mathbf{z}(t)$ where $\mathbf{z}(t)$ denotes the measured output available to the controller and $*$ denotes the convolution operator. Let

$$\mathbf{z}(t) = \mathbf{C_z}(t)\mathbf{x} \tag{50}$$

Assume that the state equation of the controller is given by

$$d\mathbf{x_K}/dt = \mathbf{A_K}(t)\mathbf{x_K} + \mathbf{B_K}(t)\mathbf{z} \tag{51}$$

$$\begin{aligned}\mathbf{u} &= \mathbf{C_K}(t)\mathbf{x_K} + \mathbf{D_K}(t)\mathbf{z}\\ &= \mathrm{C_K x_K} + \mathbf{D_K C_z x}\end{aligned} \tag{52}$$

The closed-loop system may then be represented by the augmented state-space model with $\mathbf{x_{cl}}T \underline{\Delta} [\mathbf{x}^T\ \mathbf{x_K}^T]^T$, [3, 16]:

$$d\mathbf{x_{cl}}/dt = \mathbf{A_{cl}x_{cl}} + \mathbf{B_{cl}w} + [\mathbf{I}\ \mathbf{0}]^T\mathbf{v}\underline{\Delta}\mathbf{A_{cl}x_{cl}} + \mathbf{B_{cl}w} + \zeta \tag{53}$$

$$\mathbf{A_{cl}} = \mathbf{A_p} + \mathbf{BGC} \tag{54}$$

$$\mathbf{B_{cl}} = \left[\mathbf{B_w}^T\ \mathbf{0}\right]^T \tag{55}$$

$$\mathbf{A_p} = \begin{bmatrix} \mathbf{A} & \mathbf{0} \\ \mathbf{0} & \mathbf{0} \end{bmatrix} \mathbf{B} = \begin{bmatrix} \mathbf{B_u} & \mathbf{0} \\ \mathbf{0} & \mathbf{I} \end{bmatrix} \mathbf{C} = \begin{bmatrix} \mathbf{C_z} & \mathbf{0} \\ \mathbf{0} & \mathbf{I} \end{bmatrix}$$
$$\mathbf{G} = \begin{bmatrix} \mathbf{D_K} & \mathbf{C_K} \\ \mathbf{B_K}(k) & \mathbf{A_K}(k) \end{bmatrix} \tag{56}$$

Although in the above formulation, the dimensions of $\mathbf{A}$ and $\mathbf{A_K}$ need not be the same, we shall assume theirs to be the same, for simplicity and better performance. We shall also set $\mathbf{x_K}(0) = E\{\mathbf{x}_0\} = \mathbf{0}$. Note that the covariance of the stochastic input in (53) is $E\{\zeta\zeta^T\} = [\mathbf{I}\ \mathbf{0}]^T\mathbf{\Theta}[\mathbf{I}\ \mathbf{0}]$.

$\mathbf{u}(t)$ in (52) may be further re-written as

$$\mathbf{u} == \left[\mathbf{YGV^TV} + \mathbf{YGY^TC_zY}\right]\mathbf{x_{cl}} \tag{57}$$

since

$$\mathbf{D_K} = [\mathbf{I}\ \ \mathbf{0}]\mathbf{G}[\mathbf{I}\ \ \mathbf{0}]^T \underline{\Delta} \mathbf{YGY^T} \tag{58}$$

$$\mathbf{C_K} = [\mathbf{I}\ \ \mathbf{0}]\mathbf{G}[\mathbf{0}\ \ \mathbf{I}]^T \underline{\Delta} \mathbf{YGV^T} \tag{59}$$

Substitute (57) into the cost functional:

$$\mathbf{x^T\Pi x} + \mathbf{u^TRu} = \mathbf{x_{cl}}^T\left[\mathbf{Y^T\Pi Y} + \left[\mathbf{YGV^TV} + \mathbf{YGY^TC_zY}\right]^T\mathbf{R}\left[\mathbf{YGV^TV} + \mathbf{YGY^TC_zY}\right]\mathbf{x_{cl}}\right. \tag{60}$$

and define

$$\mathbf{Q_z}\underline{\Delta}\mathbf{Y^T\Pi Y} + \left[\mathbf{YGV^TV} + \mathbf{YGY^TC_zT}\right]^T\mathbf{R}\left[\mathbf{YGV^TV} + \mathbf{YGY^TC_zY}\right]$$

Then it follows

$$\mathbf{x^T\Pi x} + \mathbf{u^TRu} = \mathbf{x_{cl}}^T\mathbf{Q_z}\mathbf{x_{cl}} \tag{61}$$

Now, substituting all these parameters for (25), we obtain

$$\begin{aligned} \mathbf{dP}(t)/dt + \mathbf{P}(t)\mathbf{A_{cl}} + \mathbf{A_{cl}}^T\mathbf{P}(t) - \mathbf{P}(t)\mathbf{B_{cl}B_{cl}}^T\mathbf{P}(t) - \gamma^{-2}\mathbf{Q_z} &= \mathbf{0} \\ \mathbf{P}(T) &= -\mathbf{Y^TQY} \end{aligned} \tag{62}$$

Since $\mathbf{A_{cl}}$ and $[\mathbf{YGV^TV} + \mathbf{YGY^TC_zY}]$ are affine in $\mathbf{G}$, (62) may be transformed into a LMI in $\mathbf{G}$.

$$\begin{bmatrix} \mathbf{\Gamma} + \mathbf{PA_{cl}} + \mathbf{A_{cl}}^T\mathbf{P} - \gamma^{-2}\mathbf{Y^T\Pi Y} & \mathbf{PB_{cl}} & \left[\mathbf{YGV^TV} + \mathbf{YGY^TC_zY}\right]^T \\ \mathbf{B_{cl}}^T\mathbf{P} & \mathbf{I} & \mathbf{0} \\ \left[\mathbf{YGV^TV} + \mathbf{YGY^TC_zY}\right] & \mathbf{0} & \gamma^2\mathbf{R}^{-1} \end{bmatrix} \geq \mathbf{0} \tag{63}$$

Hence, (62) may be solved by, at each time step t, finding $\mathbf{\Gamma}(t)$ and $\mathbf{G}(t)$ by solving the following optimization problem:

$$\min_{\mathbf{G}} \mathrm{Tr}[\boldsymbol{\Gamma}] \qquad \text{subject to (63)} \tag{64}$$

With $\boldsymbol{\Gamma}(t) = d\mathbf{P}/dt$ found, $\mathbf{P}(t\text{-}\varepsilon)$ may be computed (note that a minimum $d\mathbf{P}/dt$ would give a maximum $\mathbf{P}(t\text{-}\varepsilon)$ given $\mathbf{P}(t)$. Hence, we may iteratively work backward (numerically) toward time 0 to construct $\mathbf{P}(t)$ and $\mathbf{G}(t)$.

With $\mathbf{G}(t)$ found, the worst case $\mathbf{w}$ may be found as

$$\mathbf{w_o}(t) = -\mathbf{B_{cl}}^T\mathbf{P_o}(t)\mathbf{x_{cl}}(t), \tag{65}$$

and let

$$\mathbf{A_w} = \mathbf{A_p} + \mathbf{BGC} - \mathbf{B_{cl}}^T\mathbf{P_o} \tag{66}$$

The rest of the development is hence similar to the full information case and left to the reader.

Conceptual Algorithm 3.

Step 1: Choose a value of γ^2 and find $\mathbf{G}(t)$ and $\mathbf{P}(t)$ via (63)–(64). With $\mathbf{G}(t)$, compute $\mathbf{A_{cl}}$ and $\mathbf{B_{cl}}$ from (54) and (55), respectively.

Step 2: With the $\mathbf{P}(t)$ obtained, compute $||\mathbf{w_O}||_T^2$ with $\mathbf{x_{cl}}(0) = \mathbf{0}$. If $||\mathbf{w_O}||_T^2 = 1$, we have the optimal solution with $\mathbf{C_K}$ and $\mathbf{D_K}$ defining the optimal control law (52).

Step 3: If $||\mathbf{w_O}||_T^2 > 1$, the chosen γ^2 is too small. Set lower bound $\gamma_l^2 = \gamma^2$, increase γ^2, and go to step 1.

Step 4: If $||\mathbf{w_O}||_T^2 < 1$, the chosen γ^2 is too large. Set upper bound $\gamma_u^2 = \gamma^2$, decrease γ^2, and go to step 1.

Another variant output feedback problem is if we assume

$$\mathbf{z}(t) = \mathbf{C_z}(t)\mathbf{x} + \mathbf{D_w}\mathbf{w} + \mathbf{m} \tag{67}$$

where we assume $\mathbf{D_w}\mathbf{B_w}^T = \mathbf{0}$, and $\mathbf{m}$ a stationary zero-mean white noise process with $E\{\mathbf{mm}^T\} = \Theta$ and $E\{\mathbf{mv}^T\} = \mathbf{0}$. In this case, we may consider a strictly causal controller

$$\begin{aligned} d\mathbf{x_K}/dt &= \mathbf{A_K}(t)\mathbf{x_K} + \mathbf{B_K}(t)\mathbf{z} \\ \mathbf{u} &= \mathbf{C_K}(t)\mathbf{x_K} \qquad\qquad (\text{i.e.}\mathbf{D_K} = \mathbf{0}) \end{aligned} \tag{68}$$

Other than mathematical simplicity, choosing a strictly causal controller when the plant is *not* strictly proper may also be desirable for robustness purposes, because a strictly proper return ratio is generally more robust against un-modeled dynamics at high frequencies.

The augmented system in this case is given by:

$$\begin{aligned} d\mathbf{x_{cl}}/dt &= \mathbf{A_{cl}}\mathbf{x_{cl}} + \mathbf{B_{cl}}\mathbf{w} + \mathbf{B_\xi}\xi \\ \mathbf{B_\xi} &= [\mathbf{I}\ \ \mathbf{0}]^T + \mathbf{BG}[\mathbf{I}\ \ \mathbf{0}]^T \end{aligned} \tag{69}$$

where $\mathbf{A_{cl}}$ and $\mathbf{G}$ are given by (54)-(56) with $\mathbf{D_K} = \mathbf{0}$,

$$\xi = [\mathbf{v}^T\ \mathbf{m}^T], \text{and} E\{\xi\xi^T\} = \mathrm{diag}[\boldsymbol{\Theta}, \ominus] \tag{70}$$

where diag$[\boldsymbol{\Theta}, \Theta]$ denotes a block diagonal matrix with diagonal sub-blocks $\boldsymbol{\Theta}$ and Θ.

$$\mathbf{B_{cl}} = \left[\mathbf{B_w}^T\ \mathbf{0}\right]^T + \mathbf{BG}\left[\mathbf{D_w}^T\ \mathbf{0}\right]^T \tag{71}$$

$$\mathbf{u} = \mathbf{C_K x_K} = \mathbf{C_K V x_{cl}} \tag{72}$$

Thus

$$\begin{aligned} &||\mathbf{\Pi}^{1/2}\mathbf{x}(t)||_T^{\ 2} + ||\mathbf{R}^{1/2}\mathbf{u}(t)||_T^{\ 2} \\ &= \int^T \mathbf{x_{cl}}^T\left[\mathbf{Y}^T\mathbf{\Pi Y} + \mathbf{V}^T\mathbf{C_K}^T\mathbf{RC_K V}\right]\mathbf{x_{cl}}dt \end{aligned} \tag{73}$$

Hence, we may solve the problem by letting

$$\mathbf{Q_z} = -\left[\mathbf{Y}^T\mathbf{\Pi Y} + \mathbf{V}^T\mathbf{C_K}^T\mathbf{RC_K V}\right] \tag{74}$$

and substitute (54), (71), (56) with $\mathbf{D_K} = \mathbf{0}$, and (74) into (62) to obtain the appropriate RDE. The rest of the details and steps are similar to above $\mathbf{D_w} = \mathbf{0}$ case and left to the reader.

7. Example

We consider the example (modified) from ([11], p. 335).

$$d\mathbf{x}/dt = \mathbf{x} + \mathbf{u} + 2\mathbf{w} + \mathbf{v} \tag{75}$$

and

$$\mathbf{\Pi} = 100, \mathbf{R} = 1, \mathbf{Q} = \mathbf{0},$$

We further assume

$$E\{\mathbf{v}\} = 0, E\{\mathbf{v}\mathbf{v}^T\} = 1, E\{\mathbf{x}_0\} = 0, E\{\mathbf{x}_0\mathbf{x}_0^{\ T}\} = 1 \tag{76}$$

We shall apply Conceptual Algorithm 2 to find a linear time-invariant controller that minimizes the infinite-horizon performance index $\eta_\infty \underline{\Delta} \lim_{T \to \infty}(\eta_{ss}/T)$ of Section 5. By iterating with different values of γ^2, steady-state solutions for various relevant parameters are tabulated (**Table 1**) below:

γ^2	$-\mathbf{P_o}^\infty$	$\mathbf{X}_\infty$	$E\|\|\mathbf{w}\|\|_2^{\ 2}$	η_∞	$\eta_\infty/\mathbf{p}^2$	$\eta_\infty/(\mathbf{p}^2 + \mathbf{q}^2)$
4.5	8,44	0.118	33.6	42.04	1.25	1.22
5	5.58	0.094	11.74	17.32	1.47	1.35
8	2.035	0.063	1.03	3.065	2.97	1.51
10	1.468	0.058	0.50	1.968	3.94	1.31

Table 1.
Iterations with different values of γ^2.

where $q^2\underline{\Delta}\mathrm{Tr}\boldsymbol{\Theta} = 1$ (note also $p^2\underline{\Delta}\ E\|\mathrm{w}\|_2{}^2$). It can be seen that the desired solution occurs at about $\gamma^2 = 8$. We also see that values of $\lim_{t \to \infty} E\|\mathbf{w}\|_2$ and $\boldsymbol{\Theta}$ would determine the desirable value of γ^2 to be used. Note also that while η_∞/p^2 monotonically increases with γ^2, $\eta_\infty/(p^2 + \mathrm{Tr}\boldsymbol{\Theta})$ has a maximum with respect to γ^2 which is the worst-case combination of **w** and **v** from a power-ratio point of view. Note that $p^2 + \mathrm{Tr}\boldsymbol{\Theta}$ is a measure of the combined power of the two disturbances.

8. Conclusion

We have considered the H^{∞} state regulation problem for which the initial state of the plant is possibly nonzero, and the system is subject to worst-case nonzero-mean exogeneous input and zero-mean white noise disturbance. It should be clear that the two signals may be combined to represent a nonzero-mean disturbance which consists of both high and low frequencies components.

We note that our results here were derived from the Riccati differential equation related to the LQR, instead of like most existing approaches which are based on differential games theory. Rather than looking for a minimax saddle point as a solution, we derive our solution by maximizing $\mathbf{P}(0)$, which is directly related to the value of the cost functional concerned when the intial state is nonzero, of the Riccati differential equation via Bellman's principle of optimality. This should be familiar to anyone who is familiar with the theory of LQR. We believe that this approach offers us a new (in addition to the existing works such as [1–5], for example), and arguably also simpler, perspective to look at the solution to the H^{∞} control problem, and H^{∞} control with stochastic disturbances (compare [5, 8–10] for example). In fact, it is the recogniton that the H^{∞} and LQR problems may be solved via the same Riccati equation that enables us to develop the present simple way of incorporating a stocahstic exogenous input into H^{∞} control, in addition to nonzero initial conditions, based on the well-known stochastic LQR theory. We anticipate that this approach will enable control engineers to extend many H^{∞} results to include stochastic disturbances, capitializing on, and drawing from, the vast pool of results that have already been developed for stochastic LQR/LQG. Hence, we believe that this would be one of the main contributions of the present paper. Note that because we are still working with deteministic system matrices, established and well-known techniques such as the singular-value method or small-μ test, etc. are directly applicable to analyze the designed system [15].

We further note that the worst-case deterministic disturbance is in the form of $\mathbf{w}(t) = -\mathbf{B_w}^T\mathbf{P_0}(t)\mathbf{x}(t)$. This is useful, because if we assume the uncertain **A** coefficient matrix of the open-loop system to be $A + \boldsymbol{\Delta}$ and the absence of deterministic disturbances, then the resultant system may be modeled as $\mathbf{Ax} + \mathbf{w}$ where $\mathbf{w} = \boldsymbol{\Delta}\mathbf{x}$. It follows that if $\mathbf{P_0}\mathbf{B_w}\mathbf{B_w}^T\mathbf{P_0} \geq \boldsymbol{\Delta}^T\boldsymbol{\Delta}$ for all admissible $\boldsymbol{\Delta}$, then the controller obtained in our approach here will also guarantee the performance of the closed-loop system with performance bound (46)–(48) when the open-loop system **A** matrix is subject to perturbation $\boldsymbol{\Delta}$. Hence, here $\mathbf{B_w}$ may be construed as some free parameter/weight matrix which, together with γ^2, we may choose and vary to search for and construct a $\mathbf{P_0}$ that satisfies (35) with $d\mathbf{P}/dt = \mathbf{0}$, and $\mathbf{P_0}\mathbf{B_w}\mathbf{B_w}^T\mathbf{P_0} \geq \boldsymbol{\Delta}^T\boldsymbol{\Delta}$ for all admissible $\boldsymbol{\Delta}$ (for example, if $\boldsymbol{\Delta}$ is assumed to be an unstructured perturbation, [15], then we may assume $\sigma^2\mathbf{I} \geq \boldsymbol{\Delta}^T\boldsymbol{\Delta}$ where σ denotes the singular-value bound), thus solving *robust* the stochastic control problem with uncertain **A**-matrix). Note that if we let $\mathbf{W_B} = \mathbf{P_o}\mathbf{B_w}\mathbf{B_w}^T\mathbf{P_0}$, then for any $\mathbf{P_0}$ obtained we may find $\mathbf{B_w} = (\mathbf{P_0}^{-1}\mathbf{W_B}\mathbf{P_0}^{-1})^{1/2}$. Thus, we may substitute $\mathbf{P_0}\mathbf{B_w}\mathbf{B_w}^T\mathbf{P_0} = \sigma^2\mathbf{I}$ into (35) with $d\mathbf{P}/dt = \mathbf{0}$ to obtain

$$\mathbf{0} = -\mathbf{PA} - \mathbf{A}^T\mathbf{P} + \sigma^2\mathbf{I} - \gamma^2\mathbf{PB_u}\mathbf{R}^{-1}\mathbf{B_u}^T\mathbf{P} + \gamma^{-2}\mathbf{\Pi} \tag{77}$$

as the Riccati equation to be searched over γ^{-2} for finding the *robust performance* control problem with unstructured A-matrix perturbations satisfying $\mathbf{\Delta}^T\mathbf{\Delta} \leq \sigma^2\mathbf{I}$.

Of course, assuming that just the **A** matrix may be uncertain is unrealistic in practice. But it does suggest a possibly fruitful approach to address the "worst-case performance when subject to system matrices perturbations" problem if we are able to extend the approach to address uncertainties in the input and output matrices as well. This will be a direction of our future research.

Author details

Yung Kuan Foo[1*] and Yeng Chai Soh[2]

1 LW E&M Technologies Private Limited, Singapore

2 College of Engineering, Nanyang Technological University, Singapore

*Address all correspondence to: yungfoo@hotmail.com

References

[1] Bernstein DS, Haddad M. LQG control with an H-infinity performance bound: A Riccati equation approach. IEEE Transactions on Automatic Control. 1989;**34**(3):293-305

[2] Doyle JC, Glover K, Khargonekar PP, Francis BA. State-space solution to standard H_2 and H_{∞} control problems. IEEE Transactions on Automatic Control. 1989;**34**(8):1691-1696

[3] Dullerud GE, Paganini F. A Course in Robust Control Theory – A Convex Approach. New York, N.Y.: Springer; 2000

[4] Green M, Limebeer DJN. Linear Robust Control. Englewood Cliffs NJ: Prentice-Hall; 1995

[5] Hinrichsen D, Pritchard AJ. Stochastic H^{∞}. SIAM Journal on Control and Optimization. 1998;**36**:1505

[6] Foo YK. H^{∞} control with initial condition. IEEE Transactions on Circuits and Systems II: Express Briefs. 2006;**53**(9):867-871

[7] Li X, Xu J, Zhang H. Standard solution to mixed $H^2/H\infty$ control with regular Riccati equation. IET Control Theory and Applications. 2020;**14**(20):3643-3651

[8] Khargonekar PP, Rotea MA. Mixed $H^2/H\infty$ control: A convex optimization approach. IEEE Transactions on Automatic Control. 1991;**36**(7):824-837

[9] Saberi A, Chen BM, Sannuti P, Ly U. Simultaneous H^2/H^{∞} optimal control: The state feedback case. Automatica. 1993;**29**(6):1611

[10] Zhang W, Xie L, Chen BS. Stochastic H^2/H^{∞} Control: A Nash Game Approach. FL, USA: CRC Press; 2017

[11] Khargonekar PP, Nagpal KM, Poola KR. H^{∞} control with transient. SIAM Journal on Control and Optimization. 1991;**29**:1372-1393

[12] Burl JB. Linear Optimal Control, H^2 and H^{∞} Methods. California, USA: Addison - Wesley; 1999

[13] Brockett RW. Finite Dimensional Linear Systems. New York: Wiley; 1970

[14] Foo YK. Strengthened H^{∞} control via state feedback: A majorization approach using algebraic Riccati inequalities. IEEE Transactions on Automatic Control. 2004;**49**(5):824-827

[15] Skogestad S, Postlethwaite I. Multivariable Feedback: Analysis and Design. West Sussex, England: Wiley; 1996

[16] Skelton RE, Iwasaki T, Grigoriadis KM. A Unified Approach to Linear Control Design. New York: Taylor and Francis; 1998